Advanced Structured Materials

Volume 123

Common engineering materials reach in many applications their limits and new developments are required to fulfil increasing demands on engineering materials. The performance of materials can be increased by combining different materials to achieve better properties than a single constituent or by shaping the material or constituents in a specific structure. The interaction between material and structure may arise on different length scales, such as micro-, meso- or macroscale, and offers possible applications in quite diverse fields.

This book series addresses the fundamental relationship between materials and their structure on the overall properties (e.g. mechanical, thermal, chemical or magnetic etc.) and applications.

The topics of *Advanced Structured Materials* include but are not limited to

- classical fibre-reinforced composites (e.g. glass, carbon or Aramid reinforced plastics)
- metal matrix composites (MMCs)
- micro porous composites
- micro channel materials
- multilayered materials
- cellular materials (e.g., metallic or polymer foams, sponges, hollow sphere structures)
- porous materials
- truss structures
- nanocomposite materials
- biomaterials
- nanoporous metals
- concrete
- coated materials
- smart materials

Advanced Structured Materials is indexed in Google Scholar and Scopus.

More information about this series at http://www.springer.com/series/8611

Hamid Reza Rezaie · Hassan Beigi Rizi ·
Mojdeh Mahdi Rezaei Khamseh ·
Andreas Öchsner

A Review on Dental Materials

Springer

Hamid Reza Rezaie
Department of Engineering Materials,
Ceramic and Biomaterial Division
Iran University of Science
and Technology (IUST)
Tehran, Iran

Hassan Beigi Rizi
Department of Engineering Materials,
Ceramic and Biomaterial Division
Iran University of Science
and Technology (IUST)
Tehran, Iran

Mojdeh Mahdi Rezaei Khamseh
Department of Engineering Materials,
Ceramic and Biomaterial Division
Iran University of Science
and Technology (IUST)
Tehran, Iran

Mechanics, Surfaces and Materials
Processing (MSMP)
Arts et Metiers ParisTech
Lille, France

Andreas Öchsner
Faculty of Mechanical Engineering
Esslingen University of Applied Sciences
Esslingen am Neckar, Baden-Württemberg
Germany

ISSN 1869-8433 ISSN 1869-8441 (electronic)
Advanced Structured Materials
ISBN 978-3-030-48930-4 ISBN 978-3-030-48931-1 (eBook)
https://doi.org/10.1007/978-3-030-48931-1

This Springer imprint is published by the registered company Springer Nature Switzerland AG
The registered company address is: Gewerbestrasse 11, 6330 Cham, Switzerland

Preface

From the past decades, the healthy body has been an important issue at which taking care of orofacial organs gives a true picture of the long service life of the body, at which several measures have been employed for fulfilling demands of better speaking, chewing, tasting, swallowing, biting, and psychosocial wellbeing abilities. It is clear that the most critical features of the oral cavity are the teeth, which play a central role in that process. This organ consists of three classified tissues, the cementum, enamel, and dentin whereat bone and gingival tissue are supporting parts at which dental caries, trauma, tooth wear, and active defects lead these organs to be faced with harsh conditions. However, improvement of the mentioned defects remains a challenge in oral health for scientists and dentists. As a consequence, reviewing the principals, processing, and application of dental materials provide a basic understanding for alleviating orofacial organs problems. This book, firstly, reviews generally the primary biomaterials, and then puts an emphasis on the basic understanding of the oral cavity and the related problems. After that, reviewing wide types of applied operative dental materials like cements, ceramics, metals, polymers, and composite is carried out. Then, the third part is devoted to the understanding of toothpastes and mouthwashes. Besides, nanoparticles and 3D-printing technology for dental materials production are reviewed. In the last chapter, finally, application of the finite element method as a manner of simulation in dentistry has been reviewed.

Contents

Chapter 1
Primary Information About Biomaterials

1.1 Introduction

Biomaterials are known as any substances and materials for the aim of making devices to replace a section or a function of the body in a safe, dependable, cost-effective, and physiologically accepted manner, which is provided as synthetic or natural materials. Synthetic materials have gained attention when the first generation of materials from 1960 to 1970 was invented for usage inside a human body. Additionally, Clemson University Advisory Board suggested a definition for a biomaterial which is “a systemically and pharmacologically inert substance designed for implantation within or incorporation with living systems”. According to Fig. 1.1, a wide variety of devices and materials are applied for treatment of diseases. For instance, sutures, tooth fillings, needles, catheters, bone plates, etc. plus, in ancient civilizations artificial eyes, ears, teeth, noses and also waxes, glues, and tissues in reconstructing missing or defective parts of the human body were used [1, 2]. The application of biomaterials became practical with the arrival of an aseptic surgical technique introduced by Dr. J Lister in the 1860s. Before that, in general, surgical procedures with or without biomaterials were unsuccessful due to infections. The infections can be worsened in the presence of biomaterials that as a consequence, implants can cause a region inaccessible to immunologically competent cells of the body. The positive biocompatible implants besides modern ones were in the skeletal system. In Table 1.1, historical events of implant applications are mentioned [3].

With reference to the mentioned definitions of the biomaterials, operation, and application of biomaterials in medicine and dentistry request a broad range of knowledge and experiences with different specialties including materials science (i.e. relationship between synthetic and biological materials like metals, ceramics, polymers, composites, etc.), biology and physiology (e.g. molecular biology, anatomy, animal, and human physiology) and clinical sciences (e.g. dentistry, maxillofacial, neurosurgery, obstetrics and gynecology, ophthalmology, orthopedics, otolaryngology, plastic and reconstructive surgery, thoracic and cardiovascular surgery, veterinary medicine, surgery, etc.) [3].

H. Reza Rezaie et al., *A Review on Dental Materials*, Advanced Structured
Materials 123, https://doi.org/10.1007/978-3-030-48931-1_1

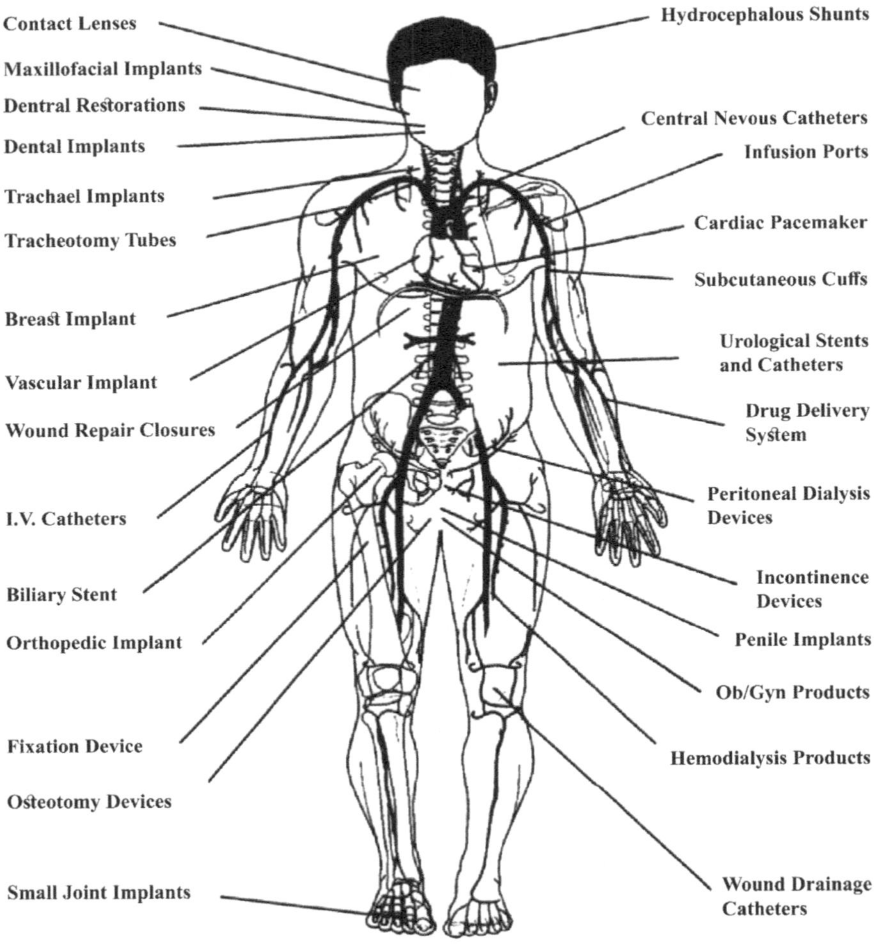

Fig. 1.1 Different aspects of implants and devices utilizing to replace or enhance the function of infections or missing tissues and organs (Reprinted from Springer Science & Business Media, Third Edition, Park & Lakes, Biomaterials: an introduction, no. 1, Copyright (2007), with permission from Springer) [1]

The selection of biomaterials for the medical application needs some terms, see Table 1.2 for a summary of them. For example, the positive response of biomaterials into the body depends on surgical routs, which means the degree of trauma imposed during implantation, sterilization methods, and so forth that are hand in hand for the outcomes [2]. Plus, some background about existing elements in human body for proposing different alloys and components are needed which are listed in Tables 1.3 and 1.4.

Table 1.1 Spots in implant developments

Year	Investigators	Developments
Late eighteenth to nineteenth century	–	Various metal devices to fix fractures; wires and pins from Fe, Au, Ag, and Pt
1860–1870	J. Lister	Aseptic surgical manner
1886	H. Hansmann	Ni-plated steel fracture plate
1893–1912	W. A. Lane	Steel screws and plates for fracture fixation
1909	A. Lambotte	Brass, Al, Ag, and Cu plate
1912	Sherman	V-steel plate, the first alloy developed exclusively for medical use
1924	A. A. Zierold	Stellite® (CoCrMo alloy), better material than Cu, Zn, steels, Mg, Fe, Ag, Au, and Al alloy
1926	M. Z. Lange	18-8sMo (2–4% Mo) stainless steel for greater corrosion resistance than 18-8 stainless steel
1926	E. W. Hey-Groves	Used carpenter's screw for femoral neck fracture
1931	M. N. Smith-Petersen	Designed first femoral neck fracture fixation nail made originally from stainless steel, later changed to Vitallium®
1936	C. S. Venable, W. G. Stuck	Vitallium®; 19 w/o Cr-9 w/o Ni stainless steel
1938	P. Wiles	First total hip replacement
1946	J. and R. Judet	First biomechanically designed hip prosthesis; first plastics used in joint replacement
1940s	M. J. Dorzee, A. Franceschetti	Acrylics for corneal replacement
1947	J. Cotton	Ti and its alloys
1952	A. B, Voorhees, A. Jaretzta, A. H. Blackmore	First blood vessel replacement made of cloth
1958	S. Furman, G. Robinson	First successful direct stimulation of the heart
1958	J. Charnley	The first use of acrylic bone cement in total hip replacements

(continued)

Table 1.1 (continued)

Year	Investigators	Developments
1960	A. Starr, M. L. Edwards	Heart valve
1970s	W. J. Kolff	Experimental total heart replacement
1990s	–	Refined implants allowing bony ingrowth
1990s	–	Controversy over silicone mammary implants
2000s	–	Tissue engineering
2000s	–	Nanoscale materials

Reprinted from Springer Science & Business Media, Third Edition, Park & Lakes, Biomaterials: an introduction, no. 11, Copyright (2007), with permission from Springer [1]

There are different considerations for the function of biomedical materials which can be divided into three main classifications. First, the biomaterials can be considered from the point of view of the problem area which is to be solved. Some areas of problems are replacement of diseased or damaged parts (e.g. the artificial hip joint, kidney dialysis machine), assist in healing (e.g. sutures, bone plates, and screws), improve function (e.g. cardiac pacemaker, intraocular lens), correct functional abnormality (e.g. cardiac pacemaker), correct cosmetic problem (e.g. augmentation mammoplasty, chin augmentation), etc. The second classification is the body consideration on a tissue level (i.e. organ or system levels). These are the heart (e.g. cardiac pacemaker, artificial heart valve, total artificial heart), lung (e.g. oxygenator machine), eye (e.g. contact lens, intraocular lens), bone (e.g. bone plate, intramedullary rod), etc. The third is for the classification of the type of materials like polymers, metals, ceramics, and composites [3]. Figure 1.2 illustrates various types of materials with respect to the general advantages and disadvantages.

Figure 1.3 represents a special materials-property chart including mechanical properties of natural and synthetic substances. Figure 1.3a illustrates the Ashby plot of the specific values of strength and stiffness which are normalized via density. The comparison between natural and engineering materials shows higher values of strength and toughness. Many natural materials have a self-healing capability against damage, on the other hand, man-made materials are still dramatically limited. According to Fig. 1.3b (dashed line indication), many natural composite materials, as exemplified by bone, have got toughness values that far exceed those for their constituents and also their homogeneous mixture which by employing extensive extrinsic toughening mechanisms, can resist incipient crack growth [4].

Table 1.2 Biomaterials selection terms for medical application

Factors	Description		
1st level material properties	Chemical/biological characteristics, chemical composition (bulk and surface)	Physical characteristic, density	Mechanical/structural characteristics, elastic modulus, Poisson's ratio, yield strength, tensile strength, compressive strength
2nd level material properties	Adhesion	Surface topology (texture and roughness)	Hardness, shear modulus, shear strength, flexural modulus, flexural strength
Specific functional requirements (based on application)	Biofunctionality (non-thrombogenic, cell adhesion, etc.), Bioinert (non-toxic, non-irritant, nonallergic, non-carcinogenic, etc.). Bioactive, biostability (resistant to corrosion, hydrolysis, oxidation and etc.), biodegradation	Form (solid, porous, coating, film, fiber, mesh, powder), geometry, coefficient of thermal expansion, electrical conductivity, color, esthetics refractive index, opacity or translucency	Stiffness or rigidity, fracture toughness, fatigue strength, creep resistance, friction and wear resistance, adhesion strength, impact strength, proof stress, abrasion resistance
Processing and fabrication	Reproducibility, quality, stabilizability, packaging, secondary processability	–	–
Characteristics of host	Tissue, organ, species, age, sex, race, health condition, activity, systemic response	–	–
Other	Medical/surgical the procedure, period of application/usage cost	–	–

Reprinted from Berlin, Germany: Springer International Publishing, Rezaie et al., Biomaterials and their applications, no. 2, Copyright (2015), with permission from Springer) [2]

1.2 Different Types of Biomaterials

1.2.1 Ceramics

Ceramics are the most inorganic refractory, polycrystalline compounds which are silicates, metallic oxides, carbides, and various refractory hydrides, sulfides, and

Table 1.3 Macroelements and the related roles in the human body

Macroelements	Roles and impacts on human body
O, C, H, N	In water and the molecular structures of proteins
Ca	Structure of bone and teeth
P	Structure of bone and teeth. Required for ATP, the energy carrier in animals
Mg	Important in bone structure. Deficiency results in tetany (muscle spasms) and can lead to a calcium deficiency
Na	The major electrolyte of blood and extracellular fluid. Required for maintenance of pH and osmotic balance
K	The major electrolyte of blood and intracellular fluid. Required for maintenance of pH and osmotic balance
Cl	The major electrolyte of blood and extracellular and intracellular fluid. Required for maintenance of pH and osmotic balance
S	The element of the essential amino acids methionine and cysteine. Included in the vitamins thiamin and biotin. As part of glutathione, it is required for detoxification. Poor growth due to reduced protein synthesis and lower glutathione levels potentially increasing oxidative or xenobiotic damage are consequences of low sulfur and methionine and/or cysteine intake

Reprinted from Materials Science and Engineering: R: Reports, Chen et al., Metallic implant biomaterials, no. 6, Copyright (2015), with permission from Elsevier [14]

selenides. Oxide ceramics are Al_2O_3, MgO, SiO_2, and so on and include metallic and nonmetallic elements. Ionic salts (NaCl, CsCl, ZnS, etc.) can cause polycrystalline aggregates, but soluble salts are not appropriate for structural biomaterials. Plus, diamond and carbonaceous structures have covalent bonds such as graphite and pyrolyzed carbons [1]. Ceramics are completely different from polymers and metals in terms of mechanical properties which are difficult to shear plastically due to the types of bonds (ionic and covalent) and very few slip systems, and also, they are not ductile and show zero creep at room temperature. As a result, ceramics are vulnerable to microcracks due to fracture elastically on the initiation of crack rather than undergoing at yield point or plastic deformation. A defect-free ceramic is powerful in subjection to tensile force. For instance, flawless glass fibers show tensile strength two times larger than high-strength steel (~7 GPa). However, higher stress at crack tips than that in the material away from the tip is proved, resulting in a stress concentration and cause weakness in materials, which the latter is attributed to the difficult prediction of the tensile strength of ceramics. Additionally, this fact is responsible for ceramics to have low tensile strength compared to compressive strength. By the way, bioceramics can be classified as follow: (1) alumina (Al_2O_3), (2) zirconia (ZrO_2), (3) carbons, (4) calcium phosphates ceramics (CPCs), (5) aluminum–calcium–phosphate (ALCAP), (6) coralline, (7) bioactive glass, and (8) bioactive glass–ceramic [3].

Alumina (Al_2O_3). Also known as aluminum oxide, is the only solid oxide from aluminum which firstly has been introduced in the 1970s. Bauxite (hydrated

Table 1.4 Micronutrients or trace elements and the related roles in the human body

Trace elements	Roles and impacts on human body
Fe	Contained in heme groups of hemoglobin and myoglobin which are required for oxygen transport in the body, as well as many other metabolic enzymes and Fe–S proteins. Part of the cytochrome p450 (CYP) family of enzymes. Anemia is the primary consequence of iron deficiency. Excess iron levels can enlarge the liver, may provoke diabetes and cardiac failure. The genetic disease hemochromatosis results from excess iron absorption. Similar symptoms can be produced through excessive transfusions required for the treatment of other diseases
Cu	Contained in enzymes of the ferroxidase system which regulates iron transport in the blood and facilitates release from storage. A structural element in the enzymes tyrosinase, cytochrome c oxidase, ascorbic acid oxidase, amine oxidases, and the antioxidant enzyme copper–zinc superoxide dismutase, among others. A copper deficiency can result in anemia from reduced ferroxidase function. Excess copper levels cause liver malfunction and are associated with genetic disorder Wilson's disease
Mn	A major component of the mitochondrial antioxidant enzyme manganese superoxide dismutase. A manganese deficiency can lead to improper bone formation and reproductive disorders. An excess of manganese can lead to poor iron absorption
I	Required for the production of thyroxine which plays an important role in metabolic rate. Deficient or excessive iodine intake can cause goiter (an enlarged thyroid gland)
Zn	Important for reproductive function due to its role in FSH (follicle-stimulating hormone) and LH (luteinizing hormone). Required for DNA binding of zinc finger proteins that regulate a variety of activities. A component of the enzyme's alcohol dehydrogenase, lactic dehydrogenase carbonic anhydrase, ribonuclease, DNA Polymerase, and the antioxidant copper-zinc superoxide dismutase. An excess of zinc may cause anemia or reduced bone formation
Se	Contained in the antioxidant enzyme glutathione peroxidase and heme oxidase. Deficiency results in oxidative membrane damage with different effects in different species. Human deficiency causes cardiomyopathy and is known as Keshan disease
Co	Contained in vitamin B12. An excess may cause cardiac failure
Mo	Contained in the enzyme xanthine oxidase. Required for the excretion of nitrogen in uric acid in birds. An excess can cause diarrhea and growth reduction
Cr	A cofactor in the regulation of sugar levels. Chromium deficiency may cause hyperglycemia (elevated blood sugar) and glucosuria (glucose in the urine). Elevated levels of some forms of chromium, such as Cr (VI), can be carcinogenic

Reprinted from Materials Science and Engineering: R: Reports, Chen et al., Metallic implant biomaterials, no. 6, Copyright (2015), with permission from Elsevier [14]

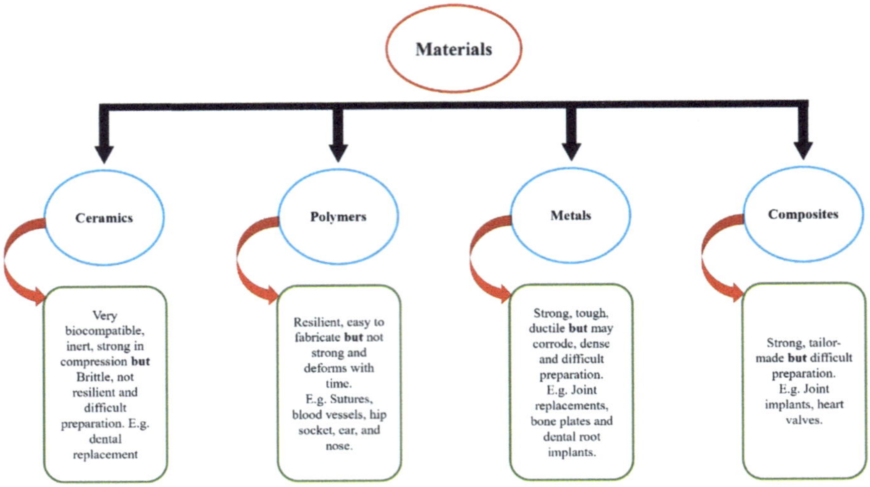

Fig. 1.2 Different materials with advantages and drawbacks based on biomedical applications [3]

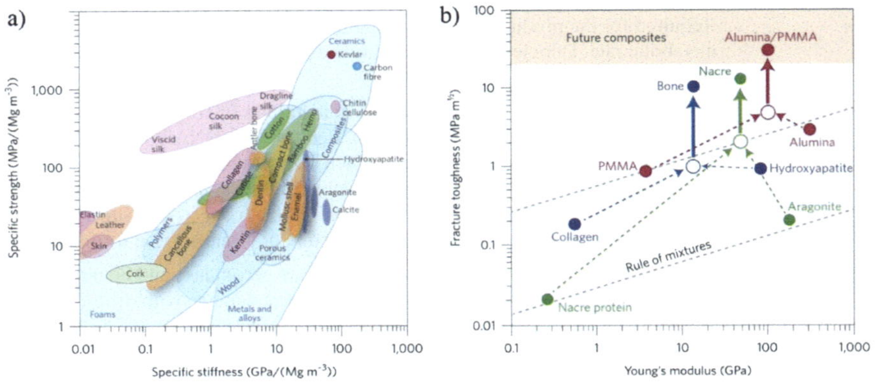

Fig. 1.3 **a** Ashby chart of strength—Young's modulus of elasticity for natural and man-made materials. Generally, it is divided into ceramics, foams, metals and their alloys, wood, polymers, porous ceramics, and composites. **b** Projections for natural and man-made materials (Reprinted from, Nature Materials, Wegst et al., Bioinspired structural materials, no. 26, Copyright (2015), with permission from Springer) [4]

aluminum oxide) and native corundum (aluminum oxide mineral or α-Al_2O_3) have been used as prime sources, which the commonly known manner for preparing is the so-called Bayer process [5]. The chemical composition of calcined alumina is Al_2O_3: 99.6, SiO_2: 0.12, Fe_2O_3: 0.03, and Na_2O: 0.04 (wt%). This material has different applications which for dentistry, according to the American Society of Testing and Materials (ASTM) of F603-78, should reach 99.5% pure alumina and less than 0.1% combination of SiO_2 and alkali oxides (mostly Na_2O). α-Al_2O_3 has a rhombohedral

Table 1.5 Mechanical and physical demands for alumina implants

Properties	Values
Flexural strength	More than 400 MPa (58,000 psi)
Elastic modulus	380 GPa (55.1×10^6 psi)
Density	3.8–3.9 g cm^{-3}

Reprinted from Springer Science & Business Media, Third Edition, Park & Lakes, Biomaterials: an introduction, no. 143, Copyright (2007), with permission from Springer [1]

crystal structure with a = 4.758 Å and c = 12.991 Å. Impurities define the color of the alumina which is attributed to the name of natural alumina for calling them sapphire or ruby. Generally, alumina is definitely a hard material with Mohs value of 9 and hardness ranges from 19.6 to 29.4 GPa. Single-crystal form of alumina effectively has been used to produce implants that were made by feeding fine alumina powders into the surface of a seed crystal and is withdrawn from an electric arc or oxy-hydrogen flame as the fused powder builds up. This method can grow the diameter of single-crystals alumina up to 0.1 mm. Another crystal formation of alumina is polycrystal at which the strength is related to the grain size and percentage of porosity. According to the ASTM standards of F603-78, flexural strength of more than 400 MPa and an elastic modulus of 380 GPa are needed (Table 1.5). One of the examples for alumina usage as a biomaterial is for total hip prostheses and combination with ultra-high-molecular-weight polyethylene (UHMWPE) socket has been outweighing in application than metals combination with UHMWPE socket [1, 3, 6].

Zirconia (ZrO_2). Also is known as zirconium oxide that has a broad application in implants (such as femoral head and acetabular cup) due to its high biocompatibility. This material is also named as "fake diamond" or "cubic zirconia" due to the same refractive index to diamond. This material is allotropic and transition from monoclinic ($a \neq b \neq c$, $\alpha = \gamma = 90° \neq \beta$) to tetragonal ($a = b \neq c$, $\alpha = \gamma = \beta = 90°$) at 1000 ~ 1200 °C and tetragonal to cubic ($a = b = c$, $\alpha = \gamma = \beta = 90°$) structure at 2370 °C. The transition from monoclinic to tetragonal phase is diffusionless, with volume reduction. The cubic structure of zirconia is from the fluorite (CaF_2) family structure that is shown in Fig. 1.4. The strength of zirconia is directly related to the phase transformation and control of grain sizes. However, zirconia is vulnerable if faced with stress in the presence of moisture which will become harsh by increasing environment working temperature. The summary of various zirconia (ceria-stabilized zirconia (CSZ), yttrium magnesium oxide-stabilized zirconia (Y-Mg-PSZ), and yttrium-stabilized zirconia (Y-TZP)) is mentioned in Table 1.6 [1].

Zircon ($ZrSiO_4$) is a mineral and gold-colored silicate of zirconium that is found in igneous and sedimentary rocks and occurring in tetragonal crystals, which are colored yellow, brown, or red based on the percentage of impurities. The preparation of zircon is started by chlorination for the aim of $ZrCl_4$ formation on a fluidized bed reactor besides of petroleum coke. Then, the second chlorination is needed for

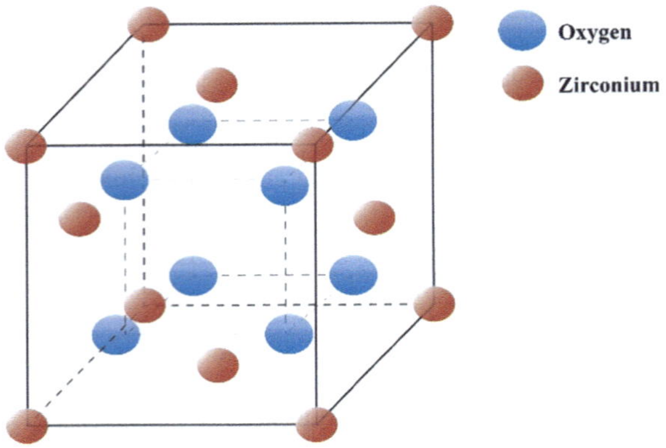

Fig. 1.4 Cubic structure of zirconia from the fluorite structure. Redrawn from Kingery [15]

Table 1.6 Properties of various zirconia

Properties	CSZ	Y-Mg-PSZ	Y-TZP
Young' s modulus (GPa)	210	210	210
Flexural strength (MPa)	200	600	950
Hardness (Vickers, HV 0.5)	1250	1250	1250
Fracture toughness (MPa $m^{1/2}$)	–	5.8	10.5
Weibull modulus	8	25	18
Density	6.1	5.85	6

Reprinted from Springer Science & Business Media, Third Edition, Park & Lakes, Biomaterials: an introduction, no. 147, Copyright (2007), with permission from Springer [1]

high-quality zirconium which is precipitated with one of the hydroxide or sulfates and, subsequently, calcined to its oxide [1].

Carbons. Besides natural allotropic crystalline forms of elemental carbon that are diamond and graphite, it has occurred in different shapes and forms in terms of crystalline which change from imperfect form including mixed amorphous, graphite-like, and diamond-like into perfectly crystalline allotropes. Table 1.7 presents a variety of carbon forms which include pyrolytic carbon, glassy or polymeric carbon, artificial graphite, carbon fibers, charcoal, vapor phase coatings, and composites. Most pure carbons are inert but solely special pyrolytic carbons provide suitable strength, fatigue resistance, wear resistance, and biodegradation resistance to operate as a long-service device. Carbons have a wide range of applications in medicine like dental implants, femoral stems, femoral condyle replacements, mechanical heart valve, ossicular replacement prosthesis, etc. [7].

Diamond, the hardest material including a network of regular tetrahedral arrays in which each carbon atom has a covalent bond with four neighbor carbon atoms forming the corners of a regular tetrahedron. There are no suitable, cost-effective

Table 1.7 Different forms of carbon

Forms	Preparation and details
Pyrolytic carbon	Produced at low or high temperature from the thermal pyrolysis of a hydrocarbon in a fluidized bed. These materials have a laminar, isotropic, granular or columnar structure and may be pure carbon or alloyed with various carbides
Glassy or polymeric carbon	Obtained from the thermal pyrolysis (~1000 °C) of selected polymers and may be monolithic, porous, or reticulated
Artificial graphite	Produced from a variety of starting materials such as petroleum or naturally occurring cokes and yield bulk structures of varying grain size, crystallite orientation, purity, porosity, strength, and particle size
Carbon fibers	Formed from spun polymeric fibers which are subsequently pyrolyzed to yield structures of unusual strength and stiffness. The properties are a function of polymer precursor and processing history. More recently, carbon fibers have been grown from the vapor phase
Charcoal	These are perhaps the oldest and most diverse materials with interesting adsorptive properties and are produced from many organic materials spanning the range from wood to coconut shells to animal bones
Vapor phase coatings	Applied, generally at reduced pressures (<1 atm) and often at low temperatures to provide a carbonaceous surface coating that ranges from amorphous to diamond-like with accompanying wide variation in thermal, mechanical, and electrical properties
Composites	Structures have been produced that utilize all of the above materials and even some other binders. Found in this group of materials are carbon fibers infiltrated and held together with pyrolytic carbon, silicon carbide, glassy carbon, PTFE, methyl methacrylate, epoxies, and petroleum pitches as well as combinations thereof. The structures may contain randomly oriented chopped fibers or long filaments oriented in random, 2, 3 and *n* dimensions

Reprinted from New York: Springer, Second Edition, Murphy et al., Handbook of biomaterial properties, no. 550, Copyright (2016), with permission from Springer [7]

manufacturing processes. So, the biomedical application of that is still challenging and needs more study.

Graphite, a layered hexagonal crystal structure at which each atom forms two single bonds and one double bond with its three nearest in-plane neighbors to form sheet-like layers of flat-six atom ring arrays. Graphite can be made in nature and generally found as isolated small scales or imperfect single-crystal precipitates in granite. It was asserted that natural graphite can be formed by metamorphosis of sedimentary carbonaceous materials or by reaction and precipitation from liquid magma which can use as a raw material for synthesis of artificial graphite. Some mechanical properties of graphite-like low strength and low wear resistance are responsible for being unsuitable for utilizing solely and without any reinforcement in implants.

Pyrolytic carbon is developed at General Atomic which could be appropriate for structural applications in high-temperature environments, gas-cooled nuclear reactors that are really in a harsh atmosphere. High-temperature-isotropic (HTI) pyrolytic carbon is derived from gas-phase nucleation and condensation of droplets formed during the pyrolysis of methane at temperature higher than 2000 °C at which densification can happen. Low-temperature-pyrolytic carbons (LTI) are prepared by the pyrolysis of other hydrocarbons same as propane and propylene at relatively lower temperatures between 1300 and 1500 °C. Pyrolytic carbon has been shown excellent blood compatibility and strength, wear resistance, and durability for long-service implants application.

Glassy or polymeric carbons, derived from a polymer via slow pyrolysis that causes a vitreous residue free of macroscopic bubbles at which the related processing is easy but has a high time-consumption. The time for producing is fluctuating from a day to one month and the temperature is more than 2000 °C at which the starting materials are phenol-formaldehyde, polyfurfuryl alcohol, polyvinyl alcohol, or oxidized polystyrene.

Carbon fibers have a wide range of applications such as incandescent electric lamp filaments and date back to the 1890s. Different manners have been proposed for fabrication of the fibers and can be classified based on their structure and the degree of crystallite orientation [7].

Calcium Phosphates Ceramics (CPCs). CPCs are the common family member of minerals including calcium cations (Ca^{2+}) and one of orthophosphate (PO_4^{3-}), metaphosphate (PO_3^-), and pyrophosphate ($P_2O_7^{4-}$) anions and in some cases contain hydrogen (H^+) ions. This material is the major portion of vertebrate bone and tooth which is widely used as repairing bone tissue and augmentation [8]. CPCs are commonly applied in medical applications in the form of cement, coatings, scaffolds, and paste which for operating properly, various properties are needed (see Table 1.8).

The CaP system and the related phases have not been explored professionally due to the key role of metastable phases, water, and kinetics which put a strain on drawing the diagram. Table 1.9 shows information about CaP phases. Among them, hydroxyapatite (HAP), tricalcium phosphate (TCP), amorphous calcium phosphates (ACPs), and biphasic calcium phosphates (BCPs) are more important and commonly used in biomedical [9].

Hydroxyapatite (HAP). One of the widely used phases of calcium phosphates is used in different biomedical grounds especially in bone regeneration due to the composition similarity with bone minerals. Also, it is the most stable and least soluble phase in which the latter would not put a halt to surface nucleating sites for the precipitation of apatite crystals in the culture medium. The mentioned properties could be affected by ionic substitutions. For instance, anionic substitution of carbonate for phosphate has been shown to increase HAP solubility, but substitution of fluoride for hydroxyapatite rise the stability. By adding cationic substitutions like magnesium to calcium phosphate instead of calcium, provide trace ions that have a favorable

Table 1.8 Properties demands for medical application of calcium phosphate ceramics (CPCs) [17–22]

Characteristics	Definitions and functions
Bioactivity	The inherent ability of a material to participate in specific biological reactions or have an effect on living tissues
Biocompatibility	The ability of a material to perform with an appropriate host response in a specific application
Bioactive fixation	Reactive surfaces form chemical bonding with bone, thus minimizing the fibrous capsule formation
Biostability	The ability of a material to maintain its properties in vivo
Crystallinity	The higher level of crystallinity prevents fast resorption (dissolution) of the bioceramic in body fluids
Interfacial stability and good adhesion	Prevent mechanical failures under load-bearing conditions
Osseointegration	Direct anchorage of an implant by the formation of bony tissue around it without growth of fibrous tissue at the bone/implant interface
Osteoconduction	Ability to provide a scaffold for the formation of new bone
Osteoinduction	The process by which osteogenesis is induced. This term means that primitive, undifferentiated and pluripotent cells are somehow stimulated to develop into the bone-forming cell lineage
Resorption	Gradual degradation overtime to replace the biomaterial with the natural host tissue
Therapeutic capabilities	Templates for the in situ delivery of drugs and growth factors at required times
Wettability	The property that indicates a material's ability to attract/repel water molecules

biological impact [10]. HAPs can be synthesized in various manners such as dry, wet, and high-temperature methods. Table 1.10 shows these methods in detail.

Tricalcium Phosphate (TCPs). It is another phase of calcium phosphate and in the stoichiometric state ($Ca_3(PO_4)_2$), it has a Ca/P ratio of 1.5 with different phases called α and β and different crystal structures. These phases are less stable in comparison with HAPs and consequently more soluble. The β-phase shows both osteoconductive and osteoinductive which can provoke apatite layer precipitation because of low interfacial energy, and also is more common than α-phase.

Amorphous Calcium Phosphates (ACPs). This phase can be produced by very fast precipitation of calcium and phosphate ions from aqueous solutions which are like a well-defined stoichiometry and marked by presenting impurities like Na^+, Cl^-, etc. ACPs have a wide range of Ca/P ration, from 1.15 to 1.67, which relies on synthesis

Table 1.9 Selected different CaP phases that are more interesting in biomedical demands

Numbers	Ca/P molar ratio	Name	Formula	pH Stability Range	Density (g/cm^3)
1	0.5	MCPM (monobasic calcium phosphate monohydrate)	$Ca(H_2PO_4)_2 \cdot H_2O$	0.0–2.0	2.22
2	1.0	DCPA (dicalcium phosphate anhydrous, Monetite)	$CaHPO_4$	2.0–5.5 (>80 °C)	2.929
3	1.0	DCPD (dibasic calcium phosphate dehydrates, brushite)	$CaHPO_4 \cdot 2H_2O$	2.0–6.0	2.319
4	1.33	OCP (octacalcium phosphate)	$Ca_8(HPO_4)_2(PO_4)_4 \cdot 5H_2O$	5.5–7.0	2.673
5	1.5	α-TCP (α-tricalcium phosphate)	$\alpha\text{-}Ca_3(PO_4)_2$	Precipitated from aqueous solutions only at T > 1125 °C	2.814[c]
6	1.5	β-TCP (β-tricalcium phosphate)	$\beta\text{-}Ca_3(PO_4)_2$	Precipitated from aqueous solutions only at T > 800°C	3.067[c]
7	1.2–2.2	ACP (amorphous calcium phosphate)	$Ca_xH_y(PO_4)_z \cdot nH_2O$ N = 3–4.5, 15–20% H_2O	~5–12[a]	–
8	1.50–1.67	CDHA (calcium deficient hydroxyapatite, CDHAp; precipitated HAp, pHA, pHAp)	$Ca_{10-x}(HPO4)_x(PO_4)_{6-x}(OH)_{2-x}$ $(0 < x < 2)$[b]	6.5–9.5	–

(continued)

Table 1.9 (continued)

Numbers	Ca/P molar ratio	Name	Formula	pH Stability Range	Density (g/cm^3)
9	1.67	HAp, or OHAp (Hydroxyapatite)	$Ca_{10}(PO_4)_6(OH)_2$	9.5–12.0	3.155
10	2.0	TTCP, or TetCP (tetracalcium phosphate, Hilgenstockite)	$Ca_4(PO_4)_2O$	Precipitated from aqueous solutions only at T > 1300 °C	3.056[c]

Reprinted from Materials Science and Engineering: C, Dorozhkin, A detailed history of calcium orthophosphates from 1770s till 1950, no. 3086, Copyright (2013), with permission from Elsevier [23]

[a]Always metastable. The composition of the precipitate depends on the composition and pH of the electrolyte solution

[b]In the case x = 1 (the boundary condition with Ca/P = 1.5), the chemical formula looks as follows: $Ca_9(HPO_4)(PO_4)_5(OH)$

[c]These compounds cannot be precipitated from aqueous solutions

Table 1.10 Different methods for producing HAPs

Category	Method	Morphology	Crystallinity degree	Phase purity	Ca/P ratio	Size	Size distribution
Dry methods	Solid-state	Diverse	Very high	Usually low	Variable	Usually micron	Wide
	Mechanochemical	Diverse	Very high	Low	Usually nonstoichiometric	Nano	Usually wide
Wet methods	Chemical precipitation	Diverse	Frequently low	Variable	Nonstoichiometric	Usually nano	Variable
	Hydrolysis	Diverse	Variable	Usually high	Stoichiometric	Variable	Variable
	Sol–gel	Diverse	Variable (usually low)	Variable	Stoichiometric	Nano	Narrow
High-temperature methods	Hydrothermal	Frequently needle-like	Very high	Usually high	Stoichiometric	Nano or micron	Usually wide
	Emulsion	Frequently needle-like	Frequently low	Variable	Nonstoichiometric	Nano	Narrow
	Sonochemical	Diverse (usually needle-like)	Variable	Usually high	Variable	Nano	Usually narrow
	Combustion	Diverse (usually irregular)	Variable	Usually high	Variable	Usually nano	Wide

(continued)

Table 1.10 (continued)

Category	Method	Morphology	Crystallinity degree	Phase purity	Ca/P ratio	Size	Size distribution
	Pyrolysis	Diverse	High	variable	Usually stoichiometric	Nanoparticles embedded in micron aggregates	Variable
	Synthesis from biogenic sources	Diverse	Variable	Usually high	Variable	Variable	Variable
	Combination procedures	Diverse (frequently needle-like)	Frequently high	Usually high	Usually stoichiometric	Usually nano	Variable

Reprinted from Berlin, Germany: Springer International Publishing, Rezaie et al., Biomaterials and their applications, no. 11, Copyright (2015), with permission from Springer [2] conditions as well as pH value. Because of their osteoinductivity, they can release calcium, phosphate, and other ions in an aqueous environment.

Biphasic Calcium Phosphate (BCPs). They are a member of a two-phase ceramics family which is the combination of low solubility and osteoconductivity of apatite in parallel with osteoinductivity of a more soluble phase such as TCPs. BCPs are formed in a wide range of shapes and morphologies like blocks, granules, cones, etc., which combining with polymers provide composites. These materials are known for bone tissue engineering scaffolds, bone regenerations, gene, and drug delivery systems [1, 2].

Aluminum-Calcium-Phosphate (ALCAP) Ceramics. Aluminum-calcium-phosphorous oxide or ALPCAP ceramics have noted a magnetic or piezoelectric property that was developed around the 1980s. This material can be provided from powders of aluminum oxide, calcium oxide, and phosphorus pentoxide. ALCAP ceramic implants provide satisfying results for biocompatibilities and gradual replacement of the ceramic materials with endogenous bone [3].

Coralline. One of the natural materials that is made by marine invertebrates is coral. This material has a porous structure which is similar to the bone structure, making it a good choice for implants application. Also, coralline could change into hydroxyapatite via a hydrothermal process. Coralline hydroxyapatite (CHA) an osteoconductive synthetic bone graft substitute material can be produced by the hydrothermal conversion of the calcium carbonate skeleton of coral to hydroxyapatite in the presence of ammonium phosphate. It can be present in the skeleton of mice, rats, guinea pigs, rabbits, dogs, etc. [11].

Bioactive Glasses. Understanding the terrible consequence of wounds sustained during the Vietnam's war in terms of amputation by professor Larry Hench caused the arrival of bioactive glasses. In the early 1970s, it was reported that the composite of Na_2O–CaO–P_2O_5–SiO_2 system with B_2O_3 and CaF_2 additives creates a strong, adherent bond with bone [12]. Bioactive glasses (i.e. amorphous silicate-based materials) are compatible with the human anatomy, bond to bone, and can lead to bone growth while dissolving over time. Formation of carbonated hydroxyapatite (HCA) layers on the glass surface when is contacting with body fluid accounts for the bonding of bioactive glasses to the bone, which has near structure and properties to the bone mineral that consequently is capable for bone formation. A five steps reaction of HCA layer on the surface of the implant is proposed, see Fig. 1.5.

The bioactive glasses have been proven that they are capable of bonding fast to the bone and being osteoinductive. However, as an example of making bioglasses, melt-processing is applied for making original bioglass that includes melting high-purity oxides (SiO_2, Na_2CO_3, $CaCO_3$, and P_2O_5) in a crucible in a furnace at 1370 °C. Platinum crucibles must be occupied to ensure that there is no contamination of the glass. Bioglass particulate is produced by pouring the melt into the water to quench, creating a frit. Then, the prepared frit is dried and grounded to the desired particle size range. Figure 1.6 depicts the compositional range of bone-bonding of bioactive glasses and also glass-ceramics which the glasses with the highest level of bioactivity

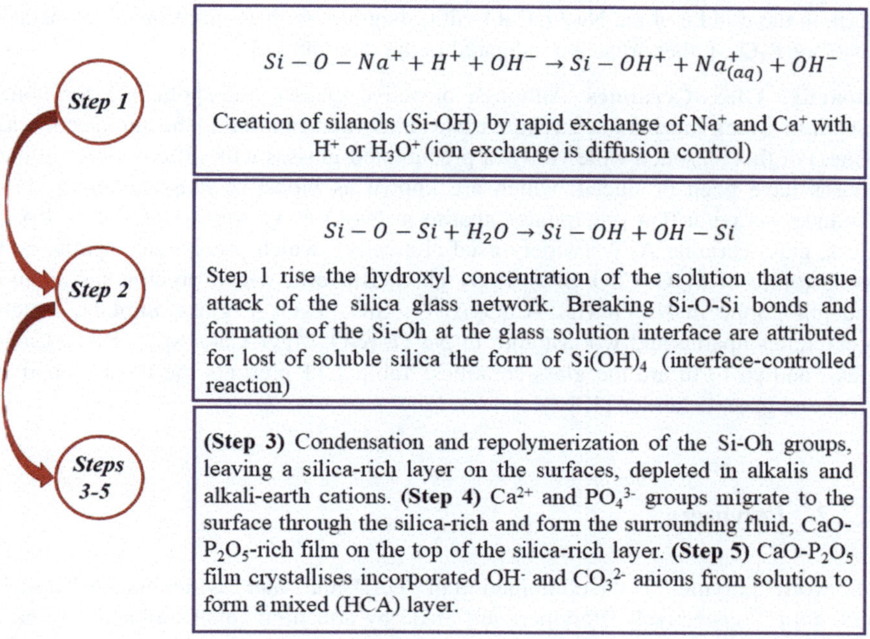

Fig. 1.5 Five steps reactions of HCA layer on the surface of the implant [13]

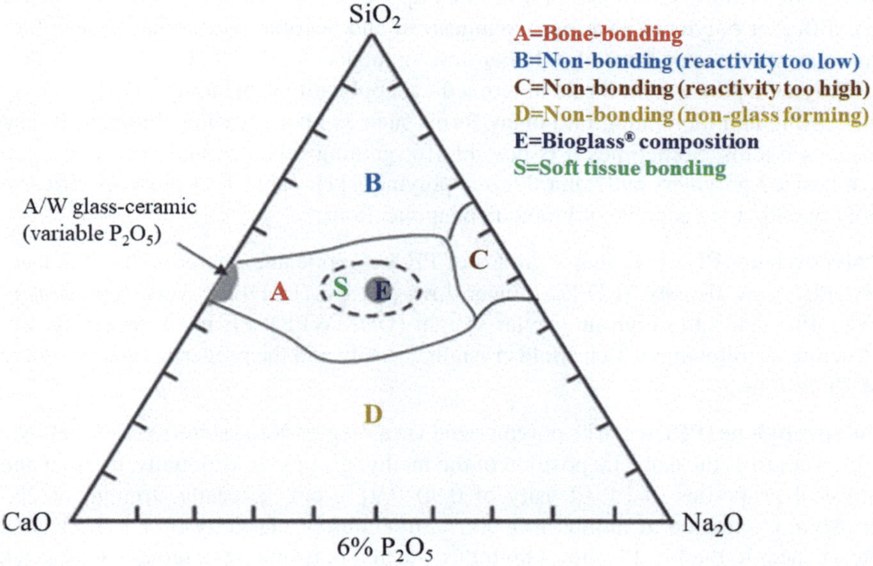

Fig. 1.6 Bon-bonding diagram based on composition. S region is Class A bioactivity where bioactive glasses bond to not only bone but also soft tissues which are gene activating. Redrawn from Jones [16]

place in the middle of the Na_2O–CaO–SiO_2 diagram, with assumption of a constant 6wt% of P_2O_5. Other areas are indicated in the diagram [13].

Bioactive Glass-Ceramics. Although bioactive glasses are applicable for bone substitutes, their mechanical strength is not suitable and similar to the human cortical bone. For this reason, a wide verity of precipitated-glasses with different crystalline phases have been produced, which are known as bioactive glass–ceramics. For instance, ceravital that precipitates apatite in Na_2O–K_2O–MgO–CaO–SiO_2–P_2O_5 glass, glass-ceramic A–W (widely used clinically), which precipitates apatite and wollastonite in MgO–CaO–SiO_2–P_2O_5 glass, Bioverit, which precipitates apatite and phlogopite in Na_2O–MgO–CaO–Al_2O_3–SiO_2–P_2O_5–F glass, implant, which precipitates apatite and wollastonite in Na_2O–K_2O–MgO–CaO–SiO_2–P_2O_5–CaF_2 glass, and so forth are the glass-ceramics. Table 1.11 presents the most bioactive glass and glass-ceramics [13].

1.2.2 Polymers

The word "polymer" is the combination of "*Poly*" and "*mer*" which means "*many*" and "*unit*", respectively. Polymers are made by attaching small molecules (mers) through primary covalent bonding in the main molecular chain backbone with C, N, O, Si, etc. The elastomeric polymers were the first occupied polymer for the aim of the synthesis of rubbers for military applications. Around WWII (World War II), different polymers have been synthesized and become widespread. Table 1.12 summarizes the spot dates of polymer developments.

Polymers play a central role in biomedical applications due to the relatively easy processing and fabricating into many forms such as fibers, textiles, films, rods, and viscous liquids. Sometimes it is possible for gaining relations and bonds between synthesized polymers and natural tissue polymers [1]. Table 1.13 presents different polymers that are capable of biomedical applications.

Polyethylene (PE). Five major grades of PE are applicable, including high density (HDPE), low density (LDPE), linear low density (LLDPE), very low density (VLDPE), and ultra-high-molecular weight (UHMWPE). PE has a repeating unit structure as following which could crystallize easily and the properties are presented in Table 1.14.

Polypropylene (PP). It can be polymerized via a Ziegler-Natta stereospecific catalyst which controls the isotactic position of the methyl groups. Additionally, thermal and physical properties of PP (density of 0.90–0.91 g cm^{-3}, tensile strength of 28–36 MPa, elongation at failure 400–900%, modulus of elasticity of 1.1–1.55 GPa, etc.) is near to the PE. PP shows high flex life and outstanding external stress-crack growth resistance and, consequently, leads to propose for finger joint and so on [3].

Polymethylmethacrylate (PMMA). Commercially, it is an amorphous material with good resistance to dilute alkalis and other inorganic solutions. Additionally, PMMA

Table 1.11 The most popular bioactive glass and glass-ceramics (wt%)

	45S5 bioglass	45S5F bioglass	45S5.4F bioglass	40S5B5 bioglass	52S4.6 bioglass	55S4.3 bioglass	KGC ceravital	KGS ceravital	KGy213 ceravital	A–W glass-ceramic	MB glass-ceramic
SiO_2	45	45	45	40	52	55	46.2	46	38	34.2	19–52
P_2O_5	6	6	6	6	6	6	–	–	–	16.3	4–24
CaO	24.5	12.25	14.7	24.5	21	19.5	20.2	33	31	44.9	3–9
$Ca(PO_3)_2$	–	–	–	–	–	–	25.5	16	13.5	–	–
CaF_2	–	12.25	9.8	–	–	–	–	–	–	0.5	–
MgO	–	–	–	–	–	–	2.9	–	–	4.6	5–15
MgF_2	–	–	–	–	–	–	–	–	–	–	–
Na_2O	24.5	24.5	24.5	24.5	21	19.5	4.8	5	4	–	3–5
K_2O	–	–	–	–	–	–	0.4	–	–	–	3–5
Al_2O_3	–	–	–	–	–	–	–	7	–	–	12–33
B_2O_3	–	–	–	5	–	–	–	–	–	–	–
Ta_2O_5/TiO_2	–	–	–	–	–	–	–	6.5	–	–	–
Structure	Glass	Glass	Glass	Glass	Glass	–	Glass-ceramic	Glass-ceramic	–	Glass-ceramic	Glass-ceramic

Reprinted from Berlin, Germany: Springer International Publishing, Rezaie et al., Biomaterials and their applications, no. 15, Copyright (2015), with permission from Springer [2]

Table 1.12 Background of some commercially important polymers since WWII

Date	Polymers
1930	Styrene-butadiene rubber
1936	Polyvinyl chloride
1936	Polychloroprene (Neoprene)
1936	Polymethylmethacrylate
1937	Polystyrene
1939	Nylon 66
1941	Polytetrafluoroethylene
1942	Unsaturated polyesters
1943	Polyethylene branched
1943	Nylon 6
1943	Silicones
1944	Polyethylene terephthalate
1947	Epoxies
1955	Polyethylene, linear
1956	Polyoxymethylene
1957	Polypropylene
1957	Polycarbonates
1964	Ionomer resins
1965	Polyimides
1970	Thermoplastic elastomers
1974	Aromatic polyamides
1980s	Ultra-high-molecular-weight polyethylene

Reprinted from Springer Science & Business Media, Third Edition, Park & Lakes, Biomaterials: an introduction, no. 174, Copyright (2007), with permission from Springer [1]

is commonly known for its exceptional light transparency around 92% transmission, high refractive index around 1.49, good weathering properties and the most biocompatible polymer. The processing of PMMA is easy and happens with conventional tools, molded, surface coating, and plasma etched with glow or corona discharge. Polymethylacrylate, polyhydroxyethyl-methacrylate (PHEMA) and polyacrylamide (PAM) are the other member of acrylic polymers family which are applicable in medicine [3].

PS and the Related Copolymers. PS or polystyrene is polymerized by free radical polymerization which is generally atactic. PS has three grades that are unmodified general-purpose PS (GPPS), high impact PS (HIPS), and PS foam. Table 1.15 provides more information about PS's grades.

Table 1.13 Polymeric material is used for biomedical and implants [2, 3]

Polymers	Applications
Polylactic and polyglycolic acid	Sutures
Polyvinylchloride (PVC)	Blood and solution bag, surgical packaging, IV sets, dialysis devices, catheter bottles, connectors, and cannulae
Polyethylene (PE)	Pharmaceutical bottle, nonwoven fabric, catheter, pouch, flexible container, and orthopedic implants
Polypropylene (PP)	Disposable syringes, blood oxygenator membrane, suture, nonwoven fabric, and artificial vascular grafts
Polymethylmethacrylate (PMMA)	Blood pump and reservoirs, membrane for blood dialyzer, implantable ocular lens, and bone cement
Polystyrene (PS)	Tissue culture flasks, roller bottles, and filter wares
Polyethylene terephthalate (PET)	Implantable suture, mesh, artificial vascular grafts, and heart valve
Polytetrafluoroethylene (PTFE)	Catheter and artificial vascular grafts
Polyurethane (PU)	The film, tubing, and components
Polyamide (Nylon)	Packaging film, catheters, sutures, and mold parts

Table 1.14 Properties of PE

Properties	Low density	High density	UHMWPE[a]	Enhanced UHMWPE[b]
Molecular weight (g mol^{-1})	$3 \sim 4 \times 10^3$	5×10^5	2×10^6	Same
Density (g cm^{-3})	0.90–0.92	0.92–0.96	0.93–0.94	Same
Tensile strength (MPa)	7.6	23–40	27 min	Higher
Elongation (%)	150	400–500	200–250	Same
Modulus of elasticity (MPa)	96–260	410–1240	Close to 2200	2200
Crystallinity (%)	50–70	70–80	Higher than HDPE	Equal or slightly higher than UHMWPE

[a]Data from ASTM F648; also. 2% deformation after 90 min of recovery subjected to 7 MPa for 24 h (D621)

[b]Same as the conventional UHMWPE (ASTM, F648). Data from *A new enhanced UHMWPE for orthopedic applications: a technical brief*, Warsaw, IN: DePuy, 1989

ABS copolymers or acrylonitrile-butadiene-styrene are made by three monomers of acrylonitrile, butadiene, and styrene, at which the ratio of them plays a central role in functional characteristics and differing physical and chemical properties [3].

Table 1.15 More information on PS's grades [3]

Name	Properties and details
GPPS	T_g: 100 °C, good transparency, lack of color, easily producible, thermal-stable, low specific gravity around 1.04–1.12 g cm^{-3}, widely used in tissue culture flasks, roller bottles, vacuum canisters, etc.
HIPS	High ductility and strength due to the rubbery modifier
PS	Processing by injection molding at 180–250 °C (processing-aids: additives like stabilizers, lubricants, and mold release agents)

Polyesters. Polyethylene-terephthalate (PET) is the most common member of polyesters which is more important because of special properties such as: highly crystalline with a high melting point of 265 °C, hydrophobicity, resistance to hydrolysis in dilute acids, and cost-effective technique for production [3].

Polyamides (Nylons). Are created by the number of carbon atoms in the repeating units, which can polymerize by step-reaction/condensation and ring-scission polymerization. Interaction of hydrogen bonding and a high degree of crystallinity is attributed to excellent fiber-forming and effective increase in strength in the fiber direction. Polyamides have different types of nylon 66, nylon 610, nylon 6, and nylon 11 which are altering in physical and mechanical properties such as density (changing between 1.05 and 1.14 g cm^{-3}), tensile strength (changing between 55 and 83 MPa), elongation (changing between 90 and 300%), softening temperatures (in the range of 26–220 °C) and so forth. [—CONH—] groups in polyamides move toward the chains one another by hydrogen bonding. As a consequence, the groups have a direct impact on the Nylons properties. As an example, T_g can be dropped by decreasing the number of [—CONH—] [3].

Fluorocarbon Polymers. Teflon or polytetrafluoroethylene (PTFE) is a commonly known fluorocarbon polymer which is made from tetrafluoroethylene under pressure with a peroxide catalyst in the presence of excess water for heat control. Other less common fluorocarbon polymers for implant applications are polytrifluorochloroethylene (PTFCE), polyvinyl fluoride (PVF), and fluorinated ethylenpropylene (FEP).

Rubbers. ASTM defines rubbers as "a material that at room temperature can be stretched repeatedly to at least twice its original length and upon release of the stress, returns immediately with force to its approximate original length". Rubbers can be categorized as silicone, natural and synthetic that are used for implant production. Natural rubbers are made by the latex of the Hevea brasiliensis tree and are proven to be compatible with blood in its pure form. The synthetic form of rubbers was progressed with the aim of natural rubber alternative. The Ziegler–Natta types of stereospecific polymerization manners made this variety possible. Another type of that is silicone rubber that was developed by Dow Corning Company. About the structure, the repeating units are dimethylsiloxane and are polymerized by condensation polymerization [1, 3]. Table 1.16 shows some properties of the rubbers' family.

Table 1.16 Properties of rubbers

Properties	Natural	Neoprene	Silicone	Urethane
Tensile strength (MPa)	7–30	20	6–7	35
Elongation at failure (%)	100–700	–	350–600	650
Hardness (Shore A durometer)	30–90	40–95	–	65
Density (g cm^{-3})	0.92	1.23	1.12–1.23	1.1–1.23

Reprinted from Springer Science & Business Media, Third Edition, Park, & Lakes, Biomaterials: an introduction, no. 191, Copyright (2007), with permission from Springer [1]

1.2.3 Metals

The first screws and bone fracture plates were made by "vanadium steel" which has been the arrival of metallic implant alloys. Metals like iron (Fe), chromium (Cr), cobalt (Co), nickel (Ni), titanium (Ti), tantalum (Ta), niobium (Nb), molybdenum (Mo), and tungsten (W) have been applied for implants production that is a little amount, can be tolerated by the human body [3]. Table 1.17 presents four main categories of metallic biomaterials with related applications based on alloying elements.

Stainless Steels (SS). Is referring to special alloy groups that are iron-based with a high percentage of chromium and varying amounts of nickel. These materials could be divided into two main groups which are the chromium and chromium–nickel

Table 1.17 Categories of metallic biomaterials and relative applications as implants

Type	Basic applications
Stainless steels	Temporary devices (fracture plates, screws, hip nails, etc.)
	Total hip replacements
Co-based alloys	Total joint replacements (wrought alloys)
	Dentistry castings
Ti-based alloys	Stem and cup of total hip replacements with CoCrMo or ceramic femoral heads
	Other permanent devices (nails, pacemakers)
NiTi	Orthodontic dental arch-wires, vascular stents, vena cava filter, Intracranial aneurysm clips, contractile artificial muscles for an artificial heart, catheter guide wires, orthopedic staples
Mg	Biodegradable orthopedic implants
Ta	Wire sutures for plastic surgery and neurosurgery
	A radiographic marker

Reprinted from Materials Science and Engineering: R: Reports, Chen et al., Metallic implant biomaterials, no. 3, Copyright (2015), with permission from Elsevier [14]

types with reference to the chemical composition. Another consideration for categorizing stainless steel is based on their characteristic microstructure: martensitic, ferric, austenitic, or duplex (austenitic and ferric). The hardness of the martensitic stainless steels causes them to be ideal for dental and surgical instruments. Ferric stainless steels have few utilizing in biomedical and austenitic one is occupied in wide non-implantable medical devices where the demands for corrosion resistance are critical [14]. Stainless steels have four types of martensitic, ferritic, austenitic, and duplex based on their characteristics of microstructure. Martensitic stainless steels are utilized in dental and surgical supplies such as bone curettes, chisels and gouges, dental burs, dental chisels, curettes, explorers, root elevators and scalers, forceps, hemostats, retractors, orthodontic pliers, and scalpels. Ferritic stainless steels are applied in very limited surgical instruments such as solid handles for instruments, guide pins, and fasteners. Austenitic stainless steels are employed in a large amount of non-implantable devices and total hip replacements. However, duplex stainless steels have not been used in the biomedical fields seriously [4]. These alloys are widely used as temporary devices in bone trauma (such as fracture plates, screws, and hip nails) or permanent implants (such as total hip replacement). The "L" in the designation shows the minute amount of carbon elements, which caused high corrosion resistance in vivo conditions [2, 14].

Cobalt-Based Alloys. These alloys have been used firstly in the 1930s. The Co–Cr–Mo alloy Vitallium was applied as a cast dental alloy and after that used for orthopedics in the 1940s. Co–Cr alloys proved excellent mechanical properties and greater than stainless steel in terms of corrosion resistance. High chromium elements are attributed to the higher form of a passive oxide layer (Cr_2O_3) in the human body environment. Other alloying contents like C, W, etc. are important which are listed in Table 1.18 according to their impacts on corrosion resistance, microstructure, and mechanical properties. About clinical applications, these alloys are utilized in dentistry and maxillofacial surgery as a partial denture, dental implants, and maxillofacial implants. Likewise, in orthopedics, they are used as fracture fixation plates and screws, and hip and knee prosthesis [14]. Technical processing has direct effects on mechanical properties of this alloy at which powder metallurgical techniques such as hot isostatic pressing (HIP) followed by forging have been used for implant usage [2].

Titanium-Based Alloys. Titanium is a low-density metallic element with around 40% less than the density of iron. With comparison to the stainless steels and cobalt-based alloys, the advantage is the specific strength but the disadvantage is the tribological behavior. Superior strength, excellent biocompatibility, and enhanced corrosion resistance are attributed to the increased application of this alloy for biomedical applications. Pure titanium progressed under an allotropic transformation at 885 °C roughly, and transforms from an HCP crystal structure (α phase) to a BCC crystal structure (β phase). Generally, according to the microstructure, titanium alloys are divided into four classes which are α-alloy, near α-alloy, α–β alloys, and β alloys. Commercially pure (CP) titanium (ASTM F67) and extra low interstitial (ELI) Ti–6Al–4V alloy (ASTM F136) are the two commonly used titanium-based implant

Table 1.18 The impacts of alloying elements on corrosion resistance, microstructure, and mechanical properties

Elements	Corrosion resistance effects	Microstructure effects	Mechanical properties changing
Cr	Cr_2O_3 to corrosion resistance	Form $Cr_{23}C_6$	Enhance wear resistance
Mo	Increase corrosion resistance	Refine grain size	Enhance solid-solution strengthening
Ni	Increase corrosion resistance	–	Enhance solid-solution strengthening, Increase castability
C	–	Form $Cr_{23}C_6$	Enhance wear resistance, Increase castability
W	Decrease corrosion resistance	Reduce shrinkage cavity, gas blowhole, and grain boundary segregation	Enhance solid-solution strengthening, Decrease corrosion fatigue strength

biomaterials. These alloys are appropriate for load enduring of implants in the environment of application. Tables 1.19 presents a deep comparison between four classes of titanium and titanium alloys.

1.2.4 Composites

Composite materials are those which involve two or more distinct constituent materials or phases on a macroscopic or microscopic size scale. In addition, composites contain one continuous phase that is the so-called matrix in which different reinforcements and matrixes cause wide verity of composites. Factors like shape, size, reinforcement properties, volume portions, and bioactivity of each part have direct or indirect impacts on the final composite properties. Composites are applied in different fields especially in biomedical science such as dental filling composites, porous implants, fibrous, and particulate composites in orthopedic implants and so on. Various reinforcements like carbon fibers, polymer fiber, ceramic particles, and glass fibers with particles have been used for enhancing composite properties [2].

Table 1.19 Advantages and disadvantages of Ti-based alloys in the aim of medical application

Ti-alloys	Advantages	Drawbacks	Biomedical utilizing
CP–Ti	Excellent corrosion resistance	Cannot be significantly strengthened by heat treatment	For non-load-bearing, corrosion-resistant applications: e.g.
	Excellent biocompatibility	Poor forge ability especially below β transit due to HCP structure	Pacemaker case
	Good weldability	Having a narrow forging temperature range	Housings for ventricular-assist devices
		Low strength at ambient temperature	The implantable fusion drug pump
		–	Dental implants
		–	Maxillofacial and craniofacial implants
		–	Screws and staple for spinal surgery
α or near α-microstructure	As above	As above	Not yet
α–β microstructure	Can be strengthened by heat treatment	–	Ti–6Al–4V and Ti–6Al–4V ELI
			Total joint replacement arthroplasty (hips and knees). Ti–6Al–7Nb
			Femoral hip stems
			Fracture fixation plates
			Spinal components
			Fasteners, nails, rods, screws, and wires Ti–3Al–2.5V Tubing and intramedullary nails
β microstructure	High hardenability	High density	–
	Good ductility and toughness, excellent forge ability and good cold rolling capability (formability) at the solution-treated condition	Low creep strength	–

(continued)

Table 1.19 (continued)

Ti-alloys	Advantages	Drawbacks	Biomedical utilizing
	Good fractural toughness	Low tensile ductility in the aged state	–
	–	low resistance to wearing	–

Reprinted from Materials Science and Engineering: R: Reports, Chen et al., Metallic implant biomaterials, no. 15, Copyright (2015), with permission from Elsevier [14]

References

1. Park, J., & Lakes, R. S. (2007). *Biomaterials: An introduction*. Springer Science & Business Media.
2. Rezaie, H. R., Bakhtiari, L., & Öchsner, A. (2015). *Biomaterials and their applications*. Berlin, Germany: Springer International Publishing.
3. Wong, J. Y., Bronzino, J. D., & Peterson, D. R. (2012). *Biomaterials: Principles and practices*. CRC Press.
4. Wegst, U. G., Bai, H., Saiz, E., Tomsia, A. P., & Ritchie, R. O. (2015). Bioinspired structural materials. *Nature Materials, 14*(1), 23.
5. LeGeros, R. Z., Lin, S., Rohanizadeh, R., Mijares, D., & LeGeros, J. P. (2003). Biphasic calcium phosphate bioceramics: Preparation, properties and applications. *Journal of Materials Science Materials in Medicine, 14*(3), 201–209.
6. Al-Sanabani, F. A., Madfa, A. A., & Al-Qudaimi, N. H. (2014). Alumina ceramic for dental applications: A review article. *Am J Mater Res, 1*(1), 26–34.
7. Murphy, W., Black, J., & Hastings, G. (Eds.). (2016). *Handbook of biomaterial properties*. New York: Springer.
8. Dorozhkin, S. V., & Epple, M. (2002). Biological and medical significance of calcium phosphates. *Angewandte Chemie International Edition, 41*(17), 3130–3146.
9. Manjubala, I., Sastry, T. P., & Kumar, R. S. (2005). Bone in-growth induced by biphasic calcium phosphate ceramic in femoral defect of dogs. *Journal of Biomaterials Applications, 19*(4), 341–360.
10. Samavedi, S., Whittington, A. R., & Goldstein, A. S. (2013). Calcium phosphate ceramics in bone tissue engineering: A review of properties and their influence on cell behavior. *Acta Biomaterialia, 9*(9), 8037–8045.
11. Damien, E., & Revell, P. A. (2004). Coralline hydroxyapatite bone graft substitute: A review of experimental studies and biomedical applications. *Journal of Applied Biomaterials and Biomechanics, 2*(2), 65–73.
12. Best, S. M., Porter, A. E., Thian, E. S., & Huang, J. (2008). Bioceramics: Past, present and for the future. *Journal of the European Ceramic Society, 28*(7), 1319–1327.
13. Kokubo, T. (Ed.). (2008). *Bioceramics and their clinical applications*. Elsevier.
14. Chen, Q., & Thouas, G. A. (2015). Metallic implant biomaterials. *Materials Science and Engineering: R: Reports, 87*, 1–57.
15. Kingery, W. D. (1976). *Introduction to ceramics* (2nd ed., pp. 449–468). Wiley.
16. Jones, J. R. (2008). Bioactive glass. In *Bioceramics and their clinical applications* (pp. 266–283). Woodhead Publishing.
17. Albrektsson, T., & Johansson, C. (2001). Osteoinduction, osteoconduction and osseointegration. *European Spine Journal, 10*(2), S96–S101.
18. Surmenev, R. A., Surmeneva, M. A., & Ivanova, A. A. (2014). Significance of calcium phosphate coatings for the enhancement of new bone osteogenesis—a review. *Acta Biomaterialia, 10*(2), 557–579.

19. Surmenev, R. A. (2012). A review of plasma-assisted methods for calcium phosphate-based coatings fabrication. *Surface & Coatings Technology, 206*(8–9), 2035–2056.
20. Xu, S., Long, J., Sim, L., Diong, C. H., & Ostrikov, K. (2005). RF plasma sputtering deposition of hydroxyapatite bioceramics: synthesis, performance, and biocompatibility. *Plasma Processes and Polymers, 2*(5), 373–390.
21. Tomsia, A. P., Launey, M. E., Lee, J. S., Mankani, M. H., Wegst, U. G., & Saiz, E. (2011). Nanotechnology approaches for better dental implants. *The International journal of oral & maxillofacial implants, 26*(Suppl), 25.
22. Mouriño, V., & Boccaccini, A. R. (2009). Bone tissue engineering therapeutics: Controlled drug delivery in three-dimensional scaffolds. *Journal of the Royal Society, Interface, 7*(43), 209–227.
23. Dorozhkin, S. V. (2013). A detailed history of calcium orthophosphates from 1770s till 1950. *Materials Science and Engineering C, 33*(6), 3085–3110.

Chapter 2
Tooth Problems and Infections

2.1 Introduction

Figure 2.1 illustrates a schematic structure of a completely developed mammalian tooth, which consists of dentin, enamel, gingiva, pulp, cementum, and bone. As can be seen, the tooth is attached to its alveolus bone socket and held in place by a thin cementum interlayer adjoining the periodontal ligament. The interior section is filled by the soft dentin-pulp complex, within that vital nutrients are active. These parts are discussed in the following subsections [1]. In addition, different dental problems including dental caries and noncarious cervical lesions (NCCLs) and related causes and effects are involved.

2.2 Anatomy of Tooth and Related Parts

According to Fig. 2.1 the most important parts are enamel, dentin, saliva, microorganism, and enzymes.

Enamel. The underlying part of the tooth structure, in terms of material, is called the enamel, which is the strongest part of the human body that is instantly mineralized tissue, with a mineral content of 96 wt% by a well-defined structure. The formation of this tissue comes from precipitation of apatite crystals (i.e. calcium, phosphate, and other element-rich materials into a protein matrix that produced by ameloblasts cells), which is known as the "calcifications" or "mineralization" process. The enamel grains are rod shapes, which make the anisotropic structure and lead to fluctuating physical properties depending on the orientation. The twisting of the enamel grains in the region of the cusp tips and incisal edges causes the strengthening. In addition, the nearest to the center and furthest from the middle parts are more homogeneous, highly mineralized, and amorphous regions. The natural color of the enamel is moderately

H. Reza Rezaie et al., *A Review on Dental Materials*, Advanced Structured Materials 123, https://doi.org/10.1007/978-3-030-48931-1_2

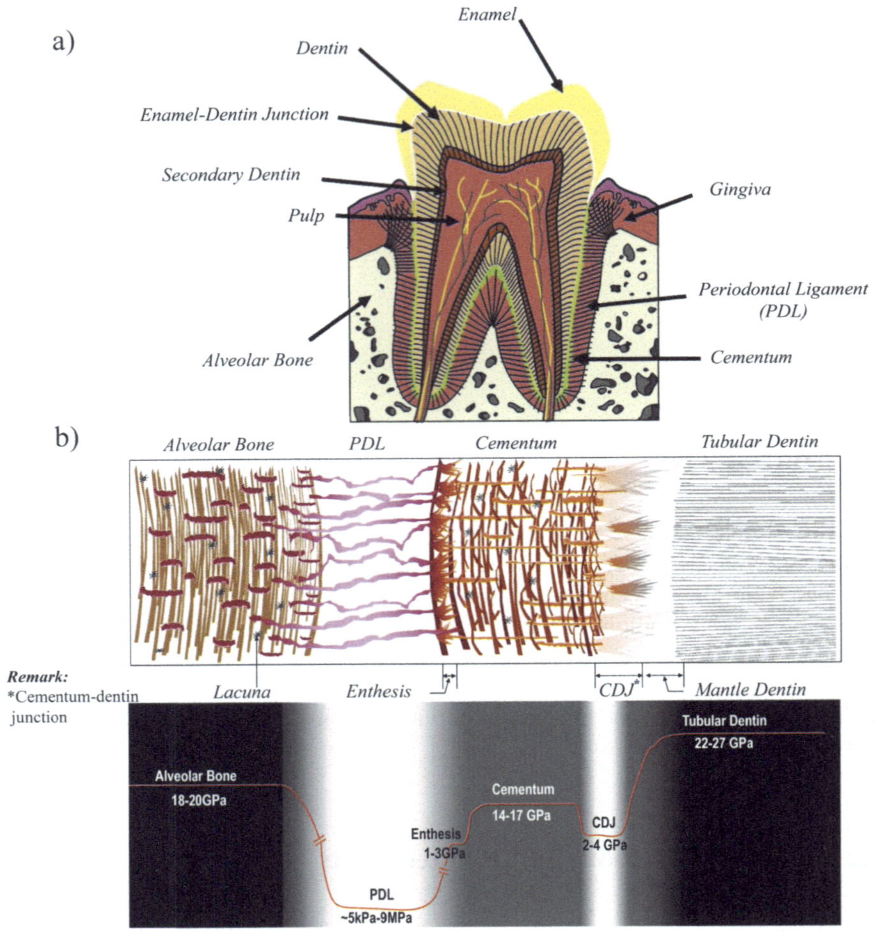

Fig. 2.1 **a** Schematic structure of a tooth (Reprinted from Acta Biomaterialia, Barani et al., Mechanics of longitudinal cracks in tooth enamel, no. 2286, Copyright (2011), with permission from Elsevier) [1], **b** an illustration of the tooth tissue and related variation of elastic modulus for each part (i.e. alveolar bone, PDL, enthesis, cementum, CDJ, and tubular dentin), (Reprinted from Biomaterials, Ho et al., The tooth attachment mechanism defined by structure, chemical composition and mechanical properties of collagen fibers in the periodontium, no. 5243 Copyright (2007), with permission from Elsevier) [2]

translucent white or whitish-blue which is presented in the incisal region and the cusp tips, where there is no significant dentine. In the thinner region of the enamel, the color of the dentin overcomes that of the enamel, which leads to make a more yellowish region. Likewise, the degree of the enamel mineralization has a direct impact on the color at which the less- and well-mineralized regions appear opaquer and relatively more translucent, respectively. It is emphatic that ions can transfer through the apatite crystal surfaces which are attributed to the shrinkage, chemical composition altering based on the local ionic conditions [1, 2].

Dentin. The dentin, a mineralized matrix collagen protein, is secreted by mesenchymal-derived odontoblasts. The structure of that is made by dentinal tubules that spread out from the pulp. The mineral content of the dentin is about 70wt%, with about 20wt% organic materials. The former is composed of crystals of apatite with the same collaboration of collagen in the bone that, however, consists of different physical arrangements. It is important to consider this fact that the odontoblastic tubules are full of fluid at which any movement of the fluid within dentinal tubes causes pain. The reason for this movement would be an osmotic pressure difference, which comes from dramatic temperature altering, during tooth cutting, and restorative operations. As mentioned above, the environment of the dentine is coated by a relatively hard enamel coat, which is separated by an enamel-dentin junction (EDJ), with several micrometers thickness. This layer is scalloped (i.e. made into a row of small curved, not a flat plane) and more in a special area that is subjected to high occlusal stress [1, 3]. Different parts of a tooth and any restorative materials should endure in a harsh condition of the oral cavity which are happening because of chewing forces during mastication, aqueous conditions, a verity of microorganism, oscillation of pH, different food substances, temperature fluctuations, and active enzymes which are discussed shortly in the following [1, 2].

Saliva. The saliva role in the human mouth is to provide a fluid layer between the sliding surfaces for controlling friction during the mastication process. Additionally, it can be useful as a buffer (i.e. buffering dietary and metabolic acids), ion storage (i.e. calcium, phosphate, and hydroxyl ions for remineralization), and antimicrobial action. Saliva exhibits lubricating properties which are related to its charged, extended macromolecules, including glycoproteins and high-molecular-weight proteins. Saliva is made up of water, organic compound, mainly proteins, and electrolytes. The members of the saliva family are three pairs of glands, the parotid, submandibular, and sublingual at which the minor glands are distributed on the tongue and cheeks [4].

Microorganism. A wide variety of microorganism inhabit the oral cavity and acids are produced by many bacteria as a by-product which lowering the intra-oral pH. As a consequence, bacteria lead to altering the dental materials, promoting hydrolysis of restorative materials [5]. This condition is attributed to natural or restorative dental material failures.

Enzymes. Enzymes that are responsible for breakdown of dental materials are not only existing in bacteria but are also existing in saliva, mineralized dentin, and the dentin fluid which include matrix metalloproteinases (MMPs) and cysteine cathepsins which are totally attributed to the hydrolytic degradation of extracellular matrix components, instantly, the adhesive hybrid layer under composite restorations [6]. However, it is clear that the collection of these conditions provides the oral cavity atmosphere with a more rapid failure process.

2.3 Dental Problems Category

The dental diseases could be divided into caries, wear-causing problems like corrosion, abrasion, abfraction, etc. From hunter-gathering to modern-day era, the most critical reason for tooth structure failure is oscillating wear via hard food substances to acidic, carbonated juices, respectively. Some side effects of modern foods may change the progress of the saliva that reduces salivary flow and leads to last decades before occurring any dental disease. According to Table 2.1, different types of dental problems by the cause and effect terms are mentioned which are covered in more details in the next following sections [3].

2.3.1 Dental Caries

Dental caries or tooth decays have been well known as microbiological, chronic diseases which are the local destruction of impressionable, susceptible hard tissue of tooth via acidic releases of fermentable bacteria of carbohydrates that are attributed to the dissolution of the dental mineral structure [7, 8]. Caries is presented not only in the crown but also in the root portions of teeth, which is known as coronal and root caries, respectively, and on pitted, fissured, and smooth dental surfaces. Likewise, it has a major impact on enamel, cementum, and dentine [9–11]. According to Fig. 2.2, tooth caries is sequential of tooth destruction stages at which altering from sub-clinical (i.e. sub-surfaces) to lesion at the molecular scale and dentin portion, respectively [12].

An equal amount of demineralization and remineralization progress put a halt to the initiation of dental caries. This progress occurred frequently every day in most people which by the time pass causes either cavitation within the dental structure or reversal of the lesion or keeping the same condition. In the caries process, endogenous bacteria commonly, mutans streptococci (streptococcus mutans and streptococcus sobrinus) and Lactobacillus spp in the dental biofilm provide weak organic acids as an additional product of metabolism of fermentable carbohydrates. Further, the by-product acid is attributed to the locally dropping pH values under a critical portion which causes in demineralization on the tooth tissues [14–17]. The cavitation will happen if the penetration of calcium, phosphate, and carbonate out of the tooth body is continuing. Calcium, phosphate, and fluoride are responsible for remineralization that fluoride propelling the diffusion of calcium and phosphate into the tooth as a catalyst and rebuilding the crystalline structure in the lesion which is consisted of fluoridated hydroxyapatite and fluorapatite that can shield more against the acid attack than the original structure [12, 14, 18]. The summary of the process is shown in Fig. 2.3.

However, a variety of factors are involved for promoting the caries progress such as poor salivary flow and composition, insufficient fluoride exposure, high numbers of cariogenic bacteria, gingival recession, immunological components, genetic terms, demands for special health care, and personal lifestyle (i.e. poor oral hygiene,

Table 2.1 Different dental diseases in terms of causes, effects, and treatments [7, 9, 10, 13, 14, 26, 31, 39, 41, 42]

	Caries	Corrosion(erosion)	Attrition	Abrasion	Abfraction
Causes	Insufficient salivary flow, Cariogenic bacteria, Inadequate fluoride exposure, Gingival recession, Immunological components, Genetic factors, Personal factors, Poverty and needs for treatment	*Intrinsic acid sources:* Vomiting, Gastroesophageal reflux, disease (GORD) *Extrinsic acid sources:* Diet, Environmental, Medications	Congenital anomalies, Psychological factors and bruxism, Parafunctional habits, Sex type, Iatrogenic	Abrasive food, Traumatizing technique of oral hygiene, Lifestyle behavior, Iatrogenic factors, Living environment	Cyclic stresses (i.e. tensile and compressive), Crack propagation during tooth flexure, Habits (e.g. bruxism)
Effects	Tooth color and appearance, Poor masticatory and facial support after treatment, Abscess forming, Enamel apatite matrix cavitation, Spreading fascial spaces and infections	Smooth silky-shining glazed surface, Occlusal erosion (cusps and concavities), Wedge-shaped lesions	A small polished facet on cusp tip, Flattering on incisal edge, Dentine exposure	Enamel, dentin or crown abrasion, Defects on anterior teeth and the occlusal surfaces, Including fatigue wear by the propagation of subsurface cracks	Wedge- or V-shaped lesions on the buccal surfaces, C-shaped with rounded floors mixed-shaped with flat, cervical and semicircular occlusal wall

(continued)

Table 2.1 (continued)

	Caries	Corrosion(erosion)	Attrition	Abrasion	Abfraction
Treatments	**Prevention**: In-office therapeutics, Patient behavior modification (i.e. diet, brushing/flossing modification), Pit-and-fissure sealants, Fluoride **Surgical intervention**: Dental filling	**Prevention**: Acid exposure dropping (paying attention to the extrinsic and intrinsic factors), Fluoride and Metal Fluoride application, Modification of acid solutions and beverages, Calcium application (e.g. calcium phosphate-containing tooth cream), Lasers application, Matrix metalloproteinases (MMPs) inhibitors agents	**Prevention**: Examination of oral patient, dental and related histories (e.g. intra-oral photographs, measuring crown height and area of the wear facet), Identifying and eliminating etiological factors (e.g. declining stress by doing sports) **Surgical intervention**: Orthodontic treatment for malocclusion disorder, Usage of ideal restorative material (e.g. metal and metal-ceramic combination)	**Prevention**: Knowledge of risk factors and modifying the lifestyle behavior (e.g. nail, pen/pencils, tacks and different hard objects biting, etc.) **Surgical intervention**: Application of toothpaste and desensitizing resin for dentin sensitivity stage, Usage of restoration materials (e.g. glass-ionomer cement, composites, etc.)	**Prevention**: Changing the patient's habits Salivary flow increasing, Monitoring the progression by the use of standardized intra-oral photographs, Study models and lesion dimension calculation, Usage of a scratch test **Surgical intervention**: Placement of restorations, Root coverage surgical procedures

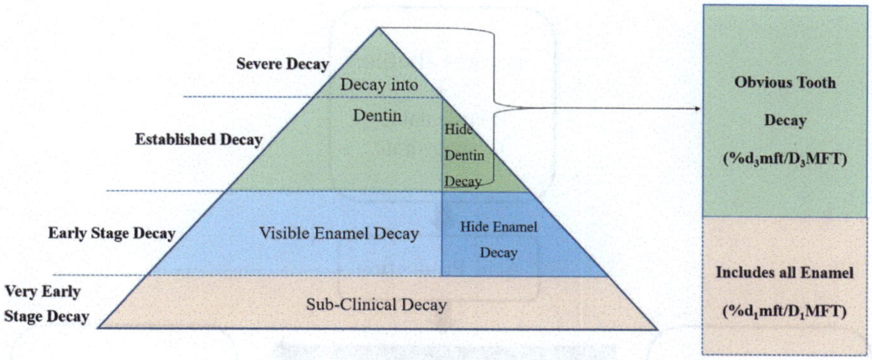

Fig. 2.2 Diagram of the iceberg metaphor for dental caries detecting the steps of caries scored at different discovering thresholds. Redrawn from Selwitz et al. [13]. The dmf and DMF are related to the primary and permanent teeth, respectively, are used to review caries statistically which is the summation of decays, missing, and filled teeth. The supporting notes relate to the diagnostic cut-off used that d_1/D_1 is attributed to enamel or dentine caries but d_3/D_3, solely, is attributed to the dentine caries

unhealthy dietary like consumption of refined carbohydrate, sugar-based foods) [7, 19–21]. Moreover, deprivation, quality of dental education, related insurance, usage of sealants, and restorative materials are affecting the orofacial health [22–24]. Figure 2.4 is a representation of the contributing factors, which are divided into the personal, oral environmental, and directly related factors to the caries development.

2.3.2 Noncarious Cervical Lesions (NCCLs)

Noncarious cervical lesions (NCCLs) or pathological tooth wear is another contributing disease that can be defined as a complex, multifactorial phenomenon with the interplay of biological, mechanical, chemical, and tribological terms, at which the portion of the disease can be determined based on muscular forces, lubricants, patient diet, behavior, and the type of restorative materials usage [26, 27]. However, dental wear can be submitted to erosion, attrition, abrasion, and abfraction terms [28–30] that are discussed specifically in the next subsections. Generally, considering each category of dental wear is incorrect as can be seen in Fig. 2.5, which illustrates their multifactorial and combinations [26].

Dental corrosion. Also known as erosion, it can be defined as the superficial loss of dental substance due to the chemical demineralization by the non-bacterial origin and acid exposure [32]. The erosion process can be divided into three main steps that are the early enamel, occlusal, and severe erosion. In the clinical aspect, early enamel erosion presents as a smooth silky-shining glazed surface. Occlusal erosion can be characterized by rounded cusps and concavities at which more progression

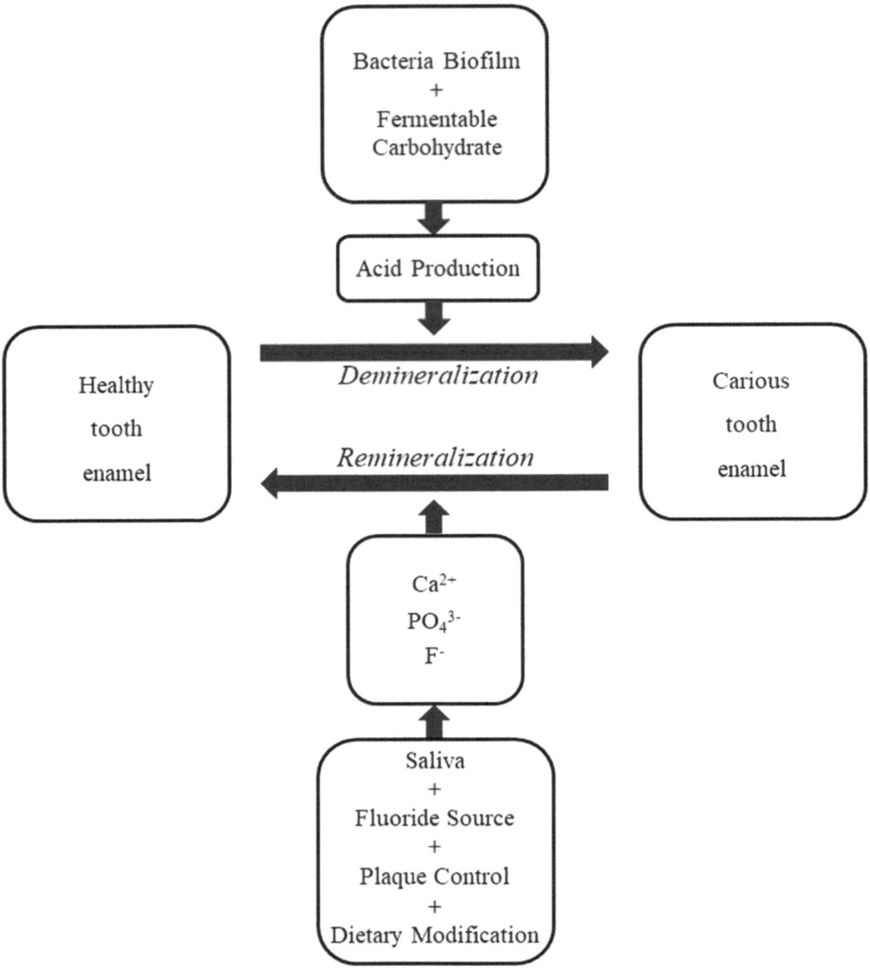

Fig. 2.3 Diagram of the remineralization and demineralization process. Redrawn from Selwitz et al. [13]

causes a distinct grooving of the cusps. In severe erosion cases, the whole occlusal or facial morphology disappears [33]. The main causes of this dental disease are multifactorial at which the important acidic sources are foods, drinks, [34] and gastric acids from reflux and regurgitation disorders [33]. However, according to Fig. 2.6, the multifactorial of the chemical, biological, and behavioral atmosphere are closely associated [32].

Acid sources are found in two main groups which can act individually or simultaneously. These groups are intrinsic and extrinsic at which the former arises from inside and the latter from outside of the body. For instance, extrinsic acid-causing

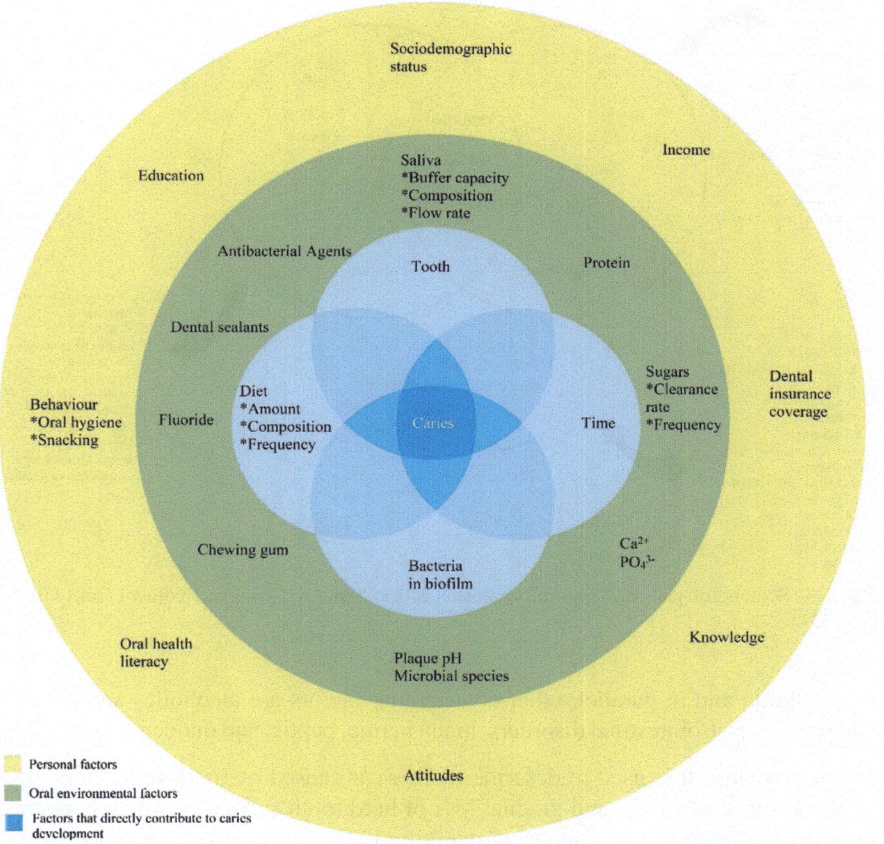

Fig. 2.4 An illustration of different kinds of factors that take place into caries development. Redrawn from Selwitz et al. [13]

erosions are soft drinks, fruit juices, pickles, fresh fruits, and yogurt, specially, that of who presents a pH lower than the critical value for enamel demineralization (Fig. 2.7, at lower pH than 4.5, acids dissolved the apatite crystals and leads to the surface lesions) which can be found in details according to the Table 2.2. Likewise, salads or low pH mouth rinses, labors who are subjected to acidic working environments and illicit drugs such as methamphetamine, cocaine, and ecstasy are playing a role [35, 36]. Intrinsic categories are the endogenous acids, which attack the tooth by vomiting, regurgitation, or reflux. Gastroesophageal reflux (GORD) diseases which are the key factor for intrinsic acid-causing is a condition where acids regurgitated back to the oral cavity from the stomach which is attributed to constant oral exposure to an acidic atmosphere. Besides, medical conditions like psychosomatic eating disorders like bulimia nervosa and anorexia nervosa are other critical causes of vomiting and

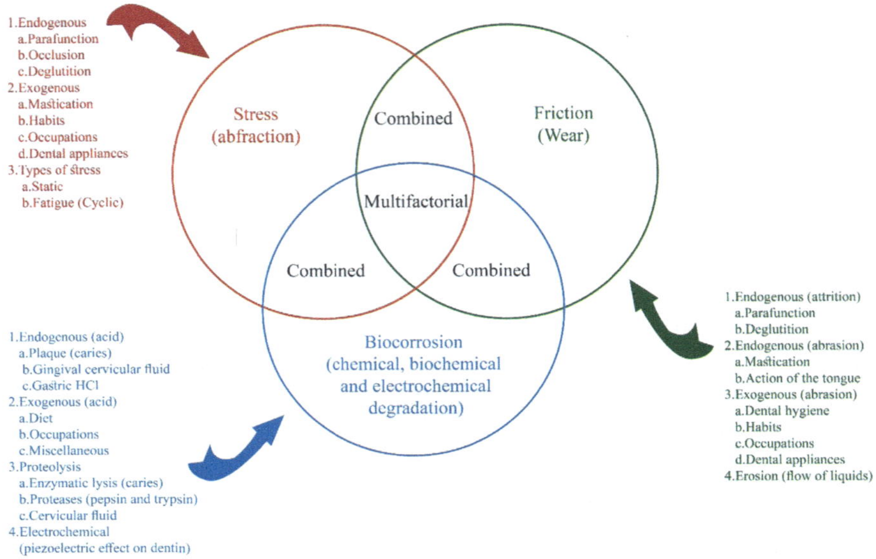

Fig. 2.5 Scheme of pathodynamic mechanisms of tooth surface lesions. Redrawn from Grippo et al. [31]

regurgitation and in parallel, other contributing factors are alcoholics and patients that present gastrointestinal disorders, hiatal hernia, peptic, and duodenal ulcers [36].

Tooth attrition. It is used to describe tooth wear caused by tooth-in-tooth contact by the absence of foods, and gradual loss of hard tooth substance which happened from occlusal contacts via opposing dentition [27]. Plus, another definition of this phenomenon adapted from every [38] is "wear caused by endogenous material like microfine particles of enamel prisms caught between two opposing teeth surfaces". In the clinical aspect, occlusal wear is attributed to the attrition that will present equal and matching wear facets on opposing teeth. The early hints are appearing by a small polished facet on a cusp tip or slight flattening on an incisal edge and subsequently, severe attrition leads to dentine exposure which could be a reason for an increase in wear rate [27]. However, there is a wide variety of involving factors that are categorized as congenital anomalies, psychological factors and bruxism, parafunctional habits, sex type, and iatrogenic. In congenital anomalies, developmental anomalies in special, amelogenesis and dentinogenesis imperfecta are attributed for raising of the tooth wear rate, which in the former, the enamel is dramatically thin and/or fragile that, in the latter, the weakness of the dentin and enamel is presented caused potential relatively separation. Psychological factors cover dental attrition sources which increase by aging. As a portion of food is masticated in the form of coarse and non-refined, the attrition problem happened due to the functionality and defined duty. Bruxism is an inevitable, non-productive behavior of gridding or clutching the

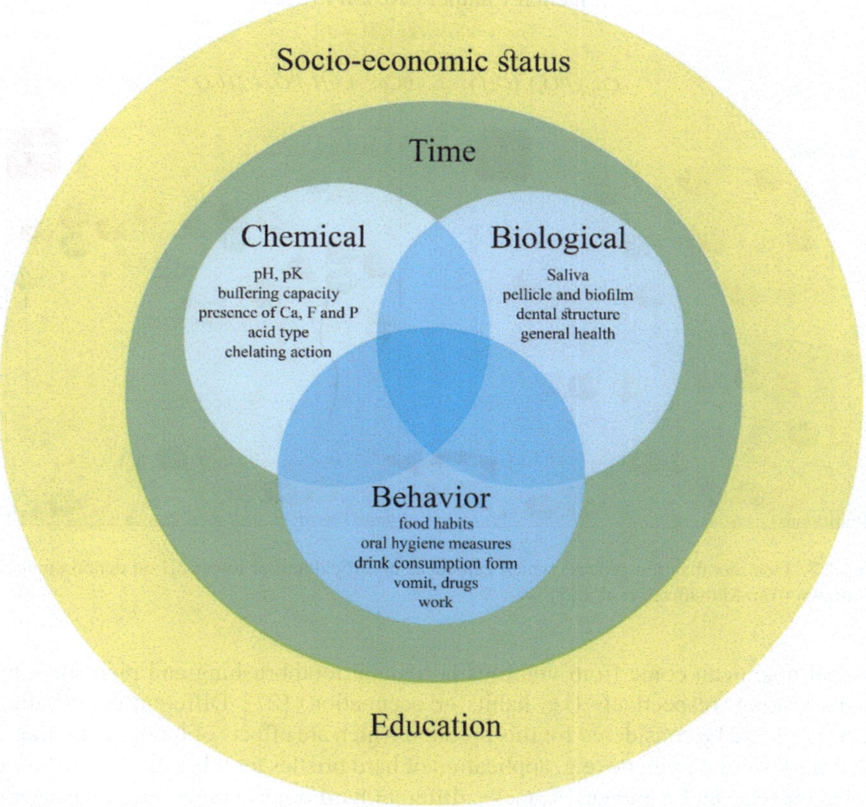

Fig. 2.6 Multifactorial conditions that account for chemical behavior and biological factors that are closely associated with the corrosion problem. Redrawn from Lussi, 2006 [32]

upper and lower teeth against each other which is destructive in nature and is alongside any stress with direct relation. Other habits besides bruxism such as pen/pencil biting, pipe smoking, and carrying hard objects between teeth causing to accelerate wear phenomenon are known as parafunctional habits. Based on gender, males are frailer than females to the attrition progress because of powerful masseter muscle activity, greater muscle fiber mass, and strong ligaments. Uncompleted occlusion due to an incorrect restoration could be uncomfortable for the patient which can force consumers to grind the teeth against each other. This iatrogenic factor speeds up in coarse porcelain restoration application [39].

Abrasion. Tooth abrasion is defined as the loss of hard tissues by the mechanical action of external substances such as food, toothbrush and toothpaste in three body wear modes [40]. The location and shape of abrasion defect are depending on foreign abrasive objects. For instance, rounded ditch on the cervical aspects of teeth and

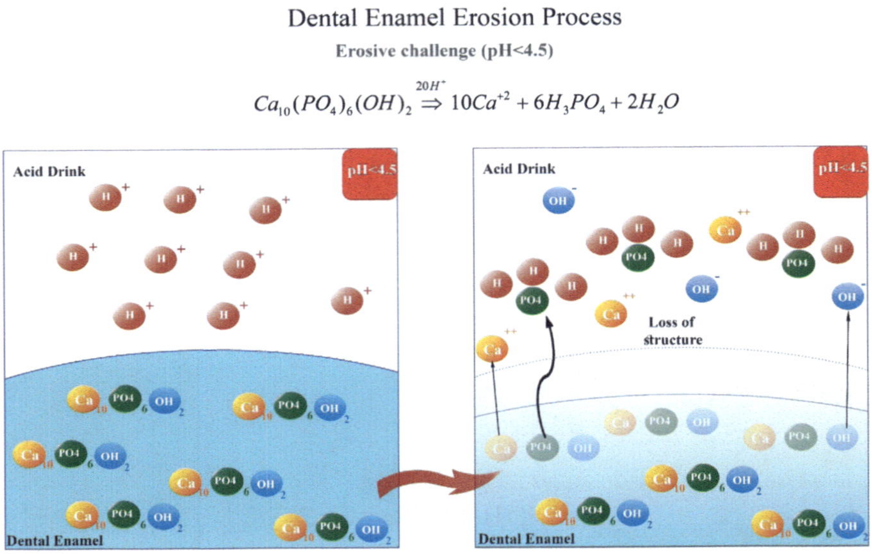

Fig. 2.7 Demineralization process which can occur in critical and/or lower pH on dental enamel. Redrawn from Mahalhaes et al. [35]

incisal notch can come from vigorous horizontal toothbrushing and pipe smoking or nail-biting, respectively (i.e. habits or occupation) [27]. Different contributing factors should be considered for this problem which are effects of food, traumatizing technique of oral hygiene (e.g. application of hard bristles toothbrush, etc.), lifestyle behavior (e.g. nail, pen/pencils, tacks, different hard objects biting etc.), iatrogenic factors (e.g. nonarticulated fillings) and living environment like desert where the hard particles like sand can get into the oral cavity [40].

Abfraction. It is another term of wear which is defined as a loss of dental hard tissue in the cervical because of crack growth during tooth flexure. Crack growth is the result of compressive and tensile stresses from mastication and malocclusion which causes wedge-shaped abfraction lesions [27]. The wear happened by a combination of bending and barreling deformations that weaken the enamel and dentin. The cyclic tension and compression may reach a fatigue limit and the crack growth progress or breakage of the tooth structure started and the opposite region comes under compressive stress, simultaneously. Some habits like bruxism cause altering the direction of forces and this leads to bending the tooth in another direction at which the stresses reverse at the related cervical area. Finally, fatigue and fracture are the outcomes [41]. Additionally, other contributing factors in wear processes like abrasion and erosion can enhance this problem [27]. Management and characterization of the lesion are totally important for controlling in an early stage. So, monitoring the related processes by clinical examination based on patient anamnesis would fulfill the demands. However, as mentioned above, this process is multifactorial at which for

Table 2.2 Acid-causing foods and beverages based on pH (Reprinted from The Journal of the American Dental Association, Marshall, T. A., Dietary assessment and counseling for dental erosion, no. 149, Copyright (2018), with permission from Elsevier) [37]

Category	Foods or beverages	Approximate pH
Beverages	Energy drinks	2.6–3.6
	Flavored waters	3.0–3.8
	Fruit juices	3.2–4.0
	Herbal teas	2.6–5.7
	Soda or pop	2.4–2.9
	Sports drinks	2.8–3.2
	wine	3.0–3.9
Fruits	Apples	3.3–4.0
	Apricots	3.3–4.8
	Blueberries	3.1–3.3
	Cherries	3.3–4.5
	Grapes	2.8–3.8
	Grapefruit	3.0–3.8
	Limes	2.0–2.8
	Mangoes	3.4–4.8
	Oranges	3.7–4.3
	Peaches	3.3–4.1
	Pears	3.5–4.6
	Pineapples	3.2–4.3
	Plums	2.8–4.3
	Pomegranates	2.9–3.2
	Raspberries	3.2–4.0
	Rhubarb	3.1–4.0
	Strawberries	3.0–3.9
Vegetables	Dill pickles	3.2–3.7
	Sauerkraut	3.3–3.6
	Tomatillos	3.8
	Tomatoes	4.3–4.9
Others	Ketchup	3.9
	Sour candies	2.5–4.3
	Vinegar	2.4–3.4

examining considering other lesion-causing, parafunction, occlusion, and oral habits are necessary. In terms of diagnosing the portion of abfraction process, the first stage happened on the buccal surfaces and are typically wedge- or V-shaped lesions with clearly defined internal and external angles. In some cases, the lesion can be in the form of C-shaped with rounded floors or mixed-shaped with the flat, cervical, and semicircular occlusal wall. In the enhanced stage, the abfraction can be deeper which is related to the other factors and NCCLs. Abfraction problems usually occurred in the adult population, with incidents arising from 3 to 17% between 20 and 70 years of age. The measurement of abfraction lesion activities needs to be considered for the treatment planning process which could be done by the use of standardized intra-oral photographs, study models, and lesion dimension calculation during the time. Plus, the assessment can be performed by the scratch test usage. The treatment strategies for this lesion are changing the patient's habits (e.g. diet, brushing technique, use of protective night guards, etc.), use of chewing gum for salivary flow increasing. Likewise, monitoring of the lesion progression, occlusal adjustments, occlusal splints, techniques to alleviate hypersensitivity, placement of restorations, and root coverage surgical procedures alongside restorations are important [26].

References

1. Barani, A., Keown, A. J., Bush, M. B., Lee, J. W., Chai, H., & Lawn, B. R. (2011). Mechanics of longitudinal cracks in tooth enamel. *Acta Biomaterialia, 7*(5), 2285–2292.
2. Ho, S. P., Marshall, S. J., Ryder, M. I., & Marshall, G. W. (2007). The tooth attachment mechanism defined by structure, chemical composition and mechanical properties of collagen fibers in the periodontium. *Biomaterials, 28*(35), 5238–5245.
3. Mount, G. J., Hume, W. R., Ngo, H. C., & Wolff, M. S. (Eds.). (2016). *Preservation and restoration of tooth structure*. Wiley.
4. Mandel, I. D. (1987). The functions of saliva. *Journal of dental research, 66*(1_suppl), 623–627.
5. Spencer, P., Ye, Q., Misra, A., Goncalves, S. D. P., & Laurence, J. S. (2014). Proteins, pathogens, and failure at the composite-tooth interface. *Journal of Dental Research, 93*(12), 1243–1249.
6. Mazzoni, A., Tjäderhane, L., Checchi, V., Di Lenarda, R., Salo, T., Tay, F. R., ... & Breschi, L. (2015). Role of dentin MMPs in caries progression and bond stability. *Journal of dental research, 94*(2), 241–251.
7. Fejerskov, O., & Kidd, E. A. M. (2003). *Dental caries: the disease and its clinical management*. Copenhagen: Denmark.
8. Marsh, P., & Martin, M. V. (1999). Oral microbiology, 4th edn. Wright.
9. Fejerskov, O. (1997). Concepts of dental caries and their consequences for understanding the disease. *Community Dentistry and Oral Epidemiology, 25*(1), 5–12.
10. Kidd, E. A. M., & Fejerskov, O. (2004). What constitutes dental caries? Histopathology of carious enamel and dentin related to the action of cariogenic biofilms. *Journal of dental research, 83*(1_suppl), 35–38.
11. Pitts, N. B. (2004). Modern concepts of caries measurement. *Journal of dental research, 83*(1_suppl), 43–47.
12. Pitts, N. (2004). ICDAS-an international system for caries detection and assessment being developed to facilitate caries epidemiology, research and appropriate clinical management. *Community Dental Health, 21,* 193–198.
13. Selwitz, R. H., Ismail, A. I., & Pitts, N. B. (2007). Dental caries. *The Lancet, 369*(9555), 51–59.
14. Featherstone, J. D. B. (2004). The continuum of dental caries—evidence for a dynamic disease process. *Journal of Dental Research, 83*(1_suppl), 39–42.
15. Fejerskov, O. (2004). Changing paradigms in concepts on dental caries: consequences for oral health care. *Caries Research, 38*(3), 182–191.
16. Scheie, A. A., & Petersen, F. C. (2004). The biofilm concept: consequences for future prophylaxis of oral diseases? *Critical Reviews in Oral Biology and Medicine, 15*(1), 4–12.
17. Caufield, P. W., & Griffen, A. L. (2000). Dental caries: an infectious and transmissible disease. *Pediatric Clinics of North America, 47*(5), 1001–1019.

18. Seow, W. K. (1998). Biological mechanisms of early childhood caries. *Community Dentistry and Oral Epidemiology, 26*(S1), 8–27.
19. Featherstone, J. D. (2003). Caries management by risk assessment: consensus statement, April 2002. *J CA Dent Assoc, 31,* 257–269.
20. Krol, D. M. (2003). Dental caries, oral health, and pediatricians. *Current problems in pediatric and adolescent health care, 33*(8), 253.
21. Hassell, T. M., & Harris, E. L. (1995). Genetic influences in caries and periodontal diseases. *Critical Reviews in Oral Biology and Medicine, 6*(4), 319–342.
22. Anderson, M. (2002). Risk assessment and epidemiology of dental caries: review of the literature. *Pediatric Dentistry, 24*(5), 377–385.
23. Thomson, W. M. (2004). Dental caries experience in older people over time: what can the large cohort studies tell us? *British Dental Journal, 196*(2), 89.
24. Winn, D. M. (2001). Tobacco use and oral disease. *Journal of Dental Education, 65*(4), 306–312.
25. Fejerskov, O., & Manji, F. (1990). Risk assessment in dental caries. *Risk Assessment in Dentistry. Chapel Hill, University of North Carolina Dental Ecology*, 215–217.
26. Nascimento, M. M., Dilbone, D. A., Pereira, P. N., Duarte, W. R., Geraldeli, S., & Delgado, A. J. (2016). Abfraction lesions: etiology, diagnosis, and treatment options. *Clinical, Cosmetic and Investigational Dentistry, 8,* 79.
27. Lee, A., He, L. H., Lyons, K., & Swain, M. V. (2012). Tooth wear and wear investigations in dentistry. *Journal of Oral Rehabilitation, 39*(3), 217–225.
28. Corica, A., & Caprioglio, A. (2014). Meta-analysis of the prevalence of tooth wear in primary dentition. *Eur J Paediatr Dent, 15*(4), 385–388.
29. Kreulen, C. M., Van't Spijker, A., Rodriguez, J. M., Bronkhorst, E. M., Creugers, N. H. J., & Bartlett, D. W. (2010). Systematic review of the prevalence of tooth wear in children and adolescents. *Caries research, 44*(2), 151–159.
30. Huysmans, M. C. D. N. J. M., Chew, H. P., & Ellwood, R. P. (2011). Clinical studies of dental erosion and erosive wear. *Caries Research, 45*(Suppl. 1), 60–68.
31. Grippo, J. O., Simring, M., & Coleman, T. A. (2012). Abfraction, abrasion, biocorrosion, and the enigma of noncarious cervical lesions: a 20-year perspective. *Journal of Esthetic and Restorative Dentistry, 24*(1), 10–23.
32. Lussi, A. (2006). Erosive tooth wear–a multifactorial condition of growing concern and increasing knowledge. In *Dental erosion* (Vol. 20, pp. 1–8). Karger Publishers.
33. Bartlett, D. (2006). Intrinsic causes of erosion. In *Dental Erosion* (Vol. 20, pp. 119–139). Karger Publishers.
34. Lussi, A., Jaeggi, T., & Zero, D. (2004). The role of diet in the aetiology of dental erosion. *Caries Research, 38*(Suppl. 1), 34–44.
35. Magalhães, A. C., Wiegand, A., Rios, D., Honório, H. M., & Buzalaf, M. A. R. (2009). Insights into preventive measures for dental erosion. *Journal of Applied Oral Science, 17*(2), 75–86.
36. Almeida e Silva, J. S., Baratieri, L. N., Araujo, E., & Widmer, N. (2011). Dental erosion: understanding this pervasive condition. *Journal of Esthetic and Restorative Dentistry, 23*(4), 205–216.
37. Marshall, T. A. (2018). Dietary assessment and counseling for dental erosion. *The Journal of the American Dental Association, 149*(2), 148–152.
38. Every, R. G. (1970). Sharpness of teeth in man and other primates. *Postilla, 143,* 1–30.
39. Jain, R., & Hegde, M. N. (2015). Dental attrition–Aetiology, diagnosis and treatment planning: a review. *J Dent Med Sci, 14,* 60–66.
40. Morozova, S. Y., Holik, M. P., Radim, M., Tomastik, M. J., Foltasova, M. L., & Harcekova, M. A. (2016). Tooth Wear-Fundamental Mechanisms and Diagnosis. *IOSR Journal of Dental and Medical Sciences, 15,* 84–91.
41. Bartlett, D. W., & Shah, P. (2006). A critical review of non-carious cervical (wear) lesions and the role of abfraction, erosion, and abrasion. *Journal of Dental Research, 85*(4), 306–312.
42. MiLoseviC, A. (2017). Abrasion: A common dental problem revisited. *Primary Dental Journal, 6*(1), 32–36.

Chapter 3
Dental Restorative Materials

Restorative dental materials are of great importance in dentistry for restoring and replacing injured or missed teeth with the purpose of simulating natural teeth functions besides providing translucency and tooth-like color shade. Restorative dental materials are produced as crowns, inlays, onlays, multi-unit fixed dental prostheses, and veneers. These materials are divided into two distinct categories, which are direct and indirect restorative materials. Direct restorative materials are fabricated and placed directly on the teeth structure, whereas indirect restorative materials are commonly used as prosthesis and are fabricated outside the oral environment. In spite of the fact that direct restorative materials including amalgam, restorative cements, and composites have been successfully used in restorative dentistry, these materials are not practically suitable to be used as multi-unit restorations [1, 2]. Introducing an ideal restorative material with favorable properties such as biocompatibility, long-term durability, high mechanical strength, appropriate fracture toughness, and tooth-like color and translucency has not been achieved yet. However, by knowing the characteristics, properties, and limitations of each restorative material it is possible to choose the best material for each special application in restorative dentistry. As a consequence, in this chapter, a wide variety of dental materials are discussed that are divided into ceramics, cements, metals, polymers, and composites.

3.1 Ceramics

3.1.1 Introduction

Ceramics are known as nonmetallic inorganic (metal-oxide or metalloid) materials exhibiting both properties of metals and nonmetals and are generally fabricated by firing at high temperatures, which presents hard and brittle behavior as well as

H. Reza Rezaie et al., *A Review on Dental Materials*, Advanced Structured Materials 123, https://doi.org/10.1007/978-3-030-48931-1_3

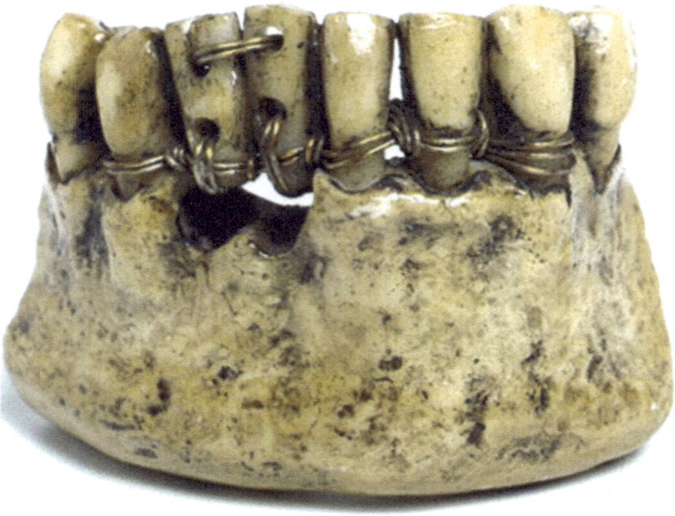

Fig. 3.1 An image of a mandible and dental prostheses held together by wires performed by Etruscans. (Reprinted from Springer Science & Business Media, Topics in Mining, Metallurgy and Materials Engineering, Bergmann C.P., & Stumpf A., Dental Ceramics Microstructure, Properties and Degradation, no. 2, Copyright (2013), with permission from Springer) [4]

poor toughness, and low thermal/electrical conductivity with various application in medicine [1, 2]. The first use of ceramics in early human history dates back to 700 years B.C. in Etruria. Archeologists have found artificial teeth made by Etruscans from ivory, animal bone, animal teeth, and human teeth sold by the poor or obtained from the dead to solve dental and mandible problems (Fig. 3.1) [1, 3, 4].

Throughout the years, various types of ceramics and fabrication methods have been used to fulfill the demands of patients and dentists in restorative dentistry. Pierre Fauchard, in 1728, established the modern dentistry by publishing the first dental book presenting restoration methods and proposing some dental instruments for better applying prosthesis. The first modern prosthesis made of hippopotamus ivory and coiled springs was exclusively prepared for George Washington who had suffered from several oral and dental problems during his lifetime [1, 4]. Chemant, a French dentist and Duchateau, a French pharmacist made the first porcelain denture in 1774 aiming to substitute ivory denture with a harder porcelain denture, which could not be used as individual teeth because of the lack of bonding that attaches teeth to the denture. The composition of denture porcelains has been developed through the years to achieve ceramic restorations with better esthetics [5]. In 1808, Fonzi, an Italian dentist, created a "terrometallic" porcelain tooth, which was held in place using a platinum pin. Planteau was a French dentist who first introduced porcelain teeth to the United States in 1817 followed by the commercial production of this product by Stockton in 1825. Pfaff in 1756 was the first one who made impressions of the mouth in Germany using plaster of Paris; however, his method was not used for many years until 1839 when the introduction of vulcanized rubber led to the use of porcelain denture [1]. Charles Land in 1886 introduced porcelain inlays, onlays, and crowns based on feldspars with high glass contents, which were highly brittle and had poor mechanical strength. Consequently, besides the high esthetics of these ceramics, they could not find their place in dentistry applications. Later, in 1962 porcelain restorations were modified by Weinstein and his colleagues with the use of high-strength metal substructures veneered with feldspathic porcelains. They came over the issue attributed to the mismatch of metal cores and porcelain veneers with different coefficients of thermal expansions. These restorations were called metal-ceramic (MC) or porcelain-fused-to-metal (PFM) restorations, which exhibit high esthetics and superior clinical performance. Metal-ceramics are already being used for total crown restorations and fixed prostheses [5]. VITA Zahnfabrik produced the first porcelain with excellent esthetics for commercial dentistry applications in 1963. Ceramco porcelains with more adaptable thermal expansion were introduced later to be used in association with a wide range of alloys. Later in 1965, McLean and Hughes developed a dental aluminous core made up of glass matrix and 40–50 wt% alumina. This core ceramic was opaque making it necessary to be utilized with porcelain veneers such as feldspathic ones. Aluminous cores have low fracture resistance while performing as molar crowns. Furthermore, Adair and Grossman in 1984 improved mechanical properties of all-ceramic dental materials with the aid of crystallization control in Dicor glass-ceramic, which contains tetrasilicic fluormica as crystalline phase. After that the machinable glass-ceramic (Dicor MGC) with almost 70 vol.% of the former crystalline phase was introduced [1]. Pressable glass-ceramic consisting of 34 vol.% leucite phases was introduced as IPS Empress (with a fracture toughness of about 1.3 MPa.m$^{1.2}$) at the beginning of the 1990s. Later, at the end of the 1990s, IPS Empress 2 with a higher fracture resistance (3.3 MPa.m$^{1.2}$) containing about 70 vol.% lithia disilicate crystalline phase was introduced. In the way of producing veneering ceramics to enhance esthetics of the lately introduced core ceramics, Duceram LFC was introduced in 1992. Duceram LFC is known as ultralow-fusing ceramic with sintering temperatures below 850 °C, high thermal expansion coefficient, self-healing behavior, and improved opalescence characteristics. In-Ceram Alumina (glass-infiltrated alumina core ceramic), In-Ceram Zirconia (glass-infiltrated zirconia alumina core ceramic), Lava (partially or fully sintered zirconia core formed by CAD/CAM process), and Cercon (partially sintered zirconia ceramic) are recently introduced dental ceramics with modified mechanical and optical properties containing high proportions of crystalline phases (Fig. 3.2) [1, 4]. These materials are very attractive, promising restorative materials that are widely used as inlays, onlays, crowns, and fixed dental prosthesis (FDPs) owing to their favorable properties such as corrosion resistance, wear resistance, low reaction tendency (i.e. with liquids, gases, alkalis, and weak acids), high chemical durability, biocompatibility, favorable esthetics, appropriate strength, and high temperature-resistance. These outstanding properties of ceramics have led them to find their place in restorative dentistry besides the ever-decreasing use of amalgam and cast metals. Until now, a wide variety of materials have been used for replacing and restoring different parts of teeth. Nowadays, there is large attention on the use

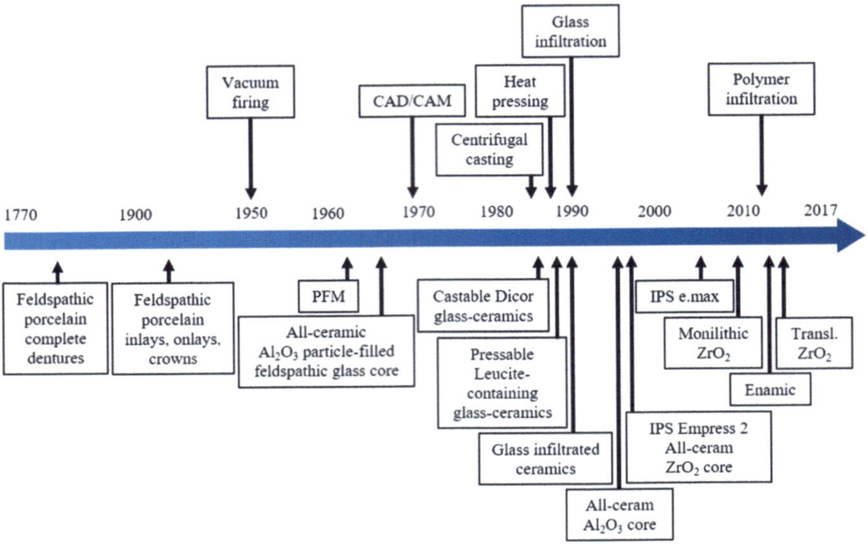

Fig. 3.2 History of the development of dental ceramics. Redrawn from Zhang et al. [5]

of high-strength all-ceramic cores instead of metal ones in prosthetic dentistry due to the esthetic limitations of the metal-ceramic restorations and high demand for providing restoratives with enhanced esthetics, mechanical properties, and clinical performance [1, 2]. However, by considering chemical and mechanical perspective, oral environment is the harshest environment in the body, and ceramics present low tensile strength and are susceptible to crack propagation; application of them in this environment faces lots of complex problems. As a matter of fact, to overcome these challenges many studies have been performed on fabricating modern dental ceramics such as shrink-free all-ceramics and castable or machinable glass-ceramics with better esthetics, long-term durability, wear resistance, and mechanical strength using advanced producing methods including heat pressing, additive manufacturing, and CAD/CAM (computer-aided design/computer-assisted manufacture) techniques [3]. In this chapter, after presenting different ceramics and metal-ceramics (MCs) with respect to orofacial applications, commonly known applied techniques such as CAD/CAM, additive manufacturing (AM), slip casting, and so on for fabricating dental ceramics products are discussed.

3.1.2 General Properties of Dental Ceramics

Most of dental ceramics comprise a crystalline phase surrounded by an amorphous matrix of silicate glass that are made up of $(SiO_4)^{4-}$ tetrahedrons linked together by sharing their corners. Crystalline silica has three different polymorphs called

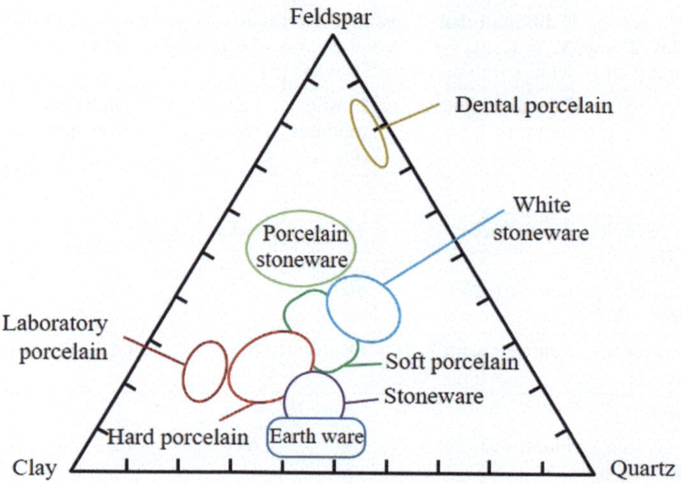

Fig. 3.3 Ternary $K_2O–Al_2O_3–SiO_2$ system determining phase fields of important feldspathic dental ceramics. (Modified and Reprinted from Journal of Non-Crystalline Solids, Zanelli et al., The vitreous phase of porcelain stoneware: composition, evolution during sintering and physical properties, no. 3253, Copyright (2011), with permission from Elsevier) [6]

quartz, cristobalite, and tridymite that are stable in different temperature ranges. In dental ceramics, quartz is a general form of crystalline silica. Since natural human teeth are ceramic with inorganic hydroxyapatite as the main component, the first restorative dental materials were composed of three major components of SiO_2, Al_2O_3, and K_2O to fabricate dental crowns (Fig. 3.3). These conventional dental ceramics were glassy materials based on silica and feldspar. However, Table 3.1 represents a comparison between different dental ceramics and tooth structure (i.e. dentin and enamel). Table 3.2 lists chemical composition of some important dental ceramics [1, 2].

3.1.3 Classification of Dental Ceramics

Dental ceramics can be classified in different ways based on their clinical application, firing temperature, microstructure, composition, crystalline, and matrix phase, fabrication method, translucency, fracture resistance, abrasiveness, etc. [1, 3].

3.1.3.1 Based on the Clinical Application

Dental ceramics are formulated and fabricated in special ways regarding their applications (e.g. crowns of anterior or posterior regions, bridges, endodontic posts or cores,

Table 3.1 Properties of different dental restorative ceramics (Reprinted from Dental Clinics of North America, Zhang, Y., & Kelly, J. R., Dental ceramics for restoration and metal veneering, no. 800, Copyright (2017), with permission from Elsevier) [5]

Material	Crystalline phase (vol.%)	Elastic modulus (GPa)	Hardnesss (GPa)	Toughness (MPa.m$^{1/2}$)	Strength (MPa)
Porcelain					
Feldspathic ceramic (Vita Mark II)	Albite (<20)	72	6.2	1.2	122
Veneer for ceramic (Lava Ceram)	Leucite (6)	80	5.2	1.1	85
Veneer for metal (d.SIGN)	Leucite/apatite (25)	68	5.9	1.1	104
Glass-Ceramic					
Mica glass-ceramic (Dicor MGC)	Fluormica (70)	69	6.0	1.2	229
Leucite glass-ceramic (IPS Empress CAD)	Leucite (35–45)	65	6.2	1.3	140
Lithium Disilicate–Ceramic					
(IPS Empress 2)	Lithium disilicate (65)	96	5.5–6.3	2.9–3.2	306–420
(IPS e.max CAD)	Lithium disilicate (70)	95	5.8	2.3	480
(IPS e.max Press)	Lithium disilicate (70)	95	5.8	2.8	400
Ceramic-Glass Interpenetrating Network					
Glass-infiltrated spine	Spinel (68)	185	–	2.5	350
Glass-infiltrated alumina	Alumina (68)	274	11.8	3.6	548
Glass-infiltrated zirconia	Zirconia-toughened alumina (67)	245	13.1	3.5	700
Polycrystalline Ceramic					
Alumina (dense, fine grain)	Alumina (>99)	372	19.6	3.1	572
Zirconia (Lava Plus)	3 mol% Y-TZP (>99)	210	14.0	4.0	1200
Zirconia (Zpex smile)	Cubic/tetragonal zirconia (>99)	210	13.4	2.4	485
Ceramic-Resin Interpenetrating Network					
Resin-infiltrated porcelain (Enamic)	Feldspathic ceramic (75)	30	1.7	1.3	159

(continued)

Table 3.1 (continued)

Material	Crystalline phase (vol.%)	Elastic modulus (GPa)	Hardnesss (GPa)	Toughness (MPa.m$^{1/2}$)	Strength (MPa)
Tooth					
Dentin	Hydroxyapatite (50)	18	0.6	3.1	34–98
Enamel	Hydroxyapatite (95)	94	3.2	0.8	12–42

Table 3.2 Chemical composition of some important dental ceramics [1, 7]

Component (wt%)	Feldspathic porcelain	IPS empress	Lithium disilicate glass-ceramic	Fluormica Glass-ceramics	IPS e.max press	IPS e.max ceram
SiO_2	52–62	59–63	57–80	56–64	57–80	45–70
Al_2O_3	11–16	19–23.5	0–5	0–2	0–5	5–22
CaO	–	0.5–3	–	–	–	1–11
Na_2O	5–7	3.5–6.5	–	–	–	4–13
K_2O	9–11	10–14	0–13	12–18	0–13	3–9
B_2O_3	–	0–1	–	–	–	–
ZnO	–	–	0–8	–	0–8	–
ZrO_2	–	–	0–8	0–5	0–8	–
BaO, Y_2O_3	–	0–1.5	–	–	–	–
SnO_2	–	–	–	–	–	–
Li_2O	–	–	11–19	–	11–19	–
F	–	–	–	4–9	–	0.1–2.5
P_2O_5	–	–	0–11	–	0–11	0.5–6.5
Sb_2O_3	–	–	–	–	–	–
CeO_2	–	0–1	–	–	–	–
TiO_2	–	0–0.5	–	–	–	–
MgO	–	–	0–5	15–20	–	–

orthodontic brackets, veneers, fixed dental prosthesis, ceramic stains, and glazes) to provide convenient performance, appropriate strength, esthetics, and durability. Two principal categories of dental ceramics classified by clinical application are (1) all-ceramic restorations for crowns, inlays, onlays, veneers of metal frameworks, anterior bridges, and fixed dental prostheses (FDPs), and (2) Porcelain-Fused-to-Metal/Metal-Ceramic (MC) crowns and fixed partial dentures (Fig. 3.4). Ceramics are also used for orthodontic brackets, dental implant abutments, and denture teeth [1, 3, 8].

Metal-ceramic crowns are composed of a metal alloy core veneered with ceramic layers, which is still widely used in restorative dentistry. These crowns present higher survival rate and fracture toughness compared to those of all-ceramics making

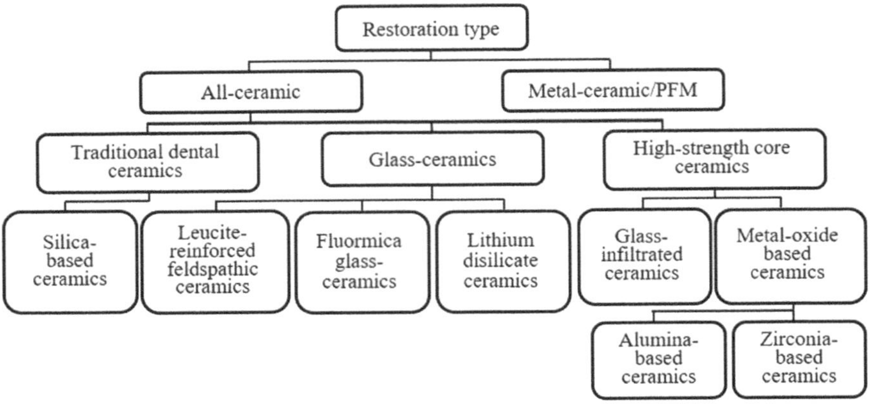

Fig. 3.4 Classification of dental ceramics by application. Redrawn from Ho et al. [8]

them suitable candidates for restoring posterior teeth. Besides this, the application of metal-ceramic restorations requires less tooth removal, which is an important factor. However, from the biologic point of view, metal-ceramics present a drawback related to metal allergy. Another disadvantage of these ceramics is their low esthetics limiting their use in anterior regions. On the other hand, all-ceramic restorations exhibit higher esthetics. All-ceramic restorations are highly crystallized reaching 99 vol.% occasionally. Size, amount, distribution, and composition of the crystalline phase affect the mechanical and optical properties of these ceramics. These ceramics can be produced using various methods including slip casting, machining, sintering, and heat pressing. The use of all-ceramics requires comparatively great removal of the tooth structure. Alumina, ceria-stabilized zirconia, glass-infiltrated alumina, glass-infiltrated magnesia-alumina spinel, glass-infiltrated alumina/zirconia, lithium disilicate glass-ceramic, yttria-stabilized zirconia are among the ceramic materials widely used as crown and bridge restorations. Table 3.3 lists some important types of dental ceramics besides their suggested applications [1, 8].

3.1.3.2 Based on Firing Temperature

Dental ceramics are classified into four different categories based on their firing temperature. In this manner, high-fusing ceramics with sintering temperatures above 1300 °C, which exhibit high-strength, chemical stability, and translucency, are used for denture teeth and fully sintered alumina and zirconia cores. Medium-fusing ceramic owing sintering temperatures ranging between 1101 °C and 1300 °C are commonly used as denture teeth. Low-fusing ceramics on the other hand, with sintering temperatures varying between 850 °C and 1100 °C are frequently applied as veneer ceramics of crowns and bridges. Ultralow-fusing ceramics with the least sintering temperature (<850 °C) are used for veneering ceramic crowns,

Table 3.3 Different applications, fabrication methods, and crystalline phases of dental ceramics [3, 7, 12]

Application	Fabrication method	Crystalline phase	Flexural strength (MPa)	Percent crystallinity	K_{IC} ($MPa.m^{1.2}$)	Products	Manufacturers
All-ceramic	Soft machining and sintering	Zirconia (3Y-TZP)	1087 ± 173	Highly crystalline	6–10	Cercon Lava IPS e.max ZirCAD In-Ceram® YZ	Dentsply 3 M Company Ivoclar Vivadent Vident
		Zirconia (cubic and tetragonal)	≈700	Highly crystalline		Zpex Smile	Tosoh Corp
		Alumina (hexagonal Al_2O_3)	≈700	Highly crystalline	4.5	In-Ceram® AL	Vident
	Hard machining	Lithium disilicate ($Li_2Si_2O_5$) Lithium silicate (Li_2SiO_3) Feldspar [(Na, K)$AlSi_3O_8$] Leucite ($KalSi_2O_6$)	262 ± 88 ≈420 122 ± 13 106 ± 17	65 >50 ≈30 ≈35	2.25–2.75 1.3	IPS e.max CAD Vita Suprinity Vitablocs® Mark II IPS Empress® CAD	Ivoclar Vivadent Vident Vident Ivoclar Vivadent
	Heat pressing	Leucite ($KalSi_2O_6$) Lithium disilicate ($Li_2Si_2O_5$) Fluorapatite [$Ca_5(PO_4)_3F$]	106 ± 17 306 ± 29	35 65		IPS Empress® IPS e.max Press IPS e.max ZirPress	Ivoclar Vivadent Ivoclar Vivadent Ivoclar Vivadent

(continued)

Table 3.3 (continued)

Application	Fabrication method	Crystalline phase	Flexural strength (MPa)	Percent crystallinity	K_{IC} (MPa.m$^{1.2}$)	Products	Manufacturers
	Sintering	Leucite ($KalSi_2O_6$) Fluorapatite [$Ca_5(PO_4)_3F$]	104 ≈80	35–40 10		IPS Empress® layering IPS e.max Ceram layering	Ivoclar Vivadent Ivoclar Vivadent
	Dry-pressing and Sintering	Alumina (Al_2O_3)	607 ± 73	Highly crystalline		Procera® All-Ceram	Nobel Biocare
	Slip-casting (soft machining) and glass-infiltrating	Alumina (Al_2O_3) Spinel ($MgAl_2O_4$) Zirconia (ZrO_2)	594 ± 52 378 ± 65 630 ± 58	67–68 65–68 67	3.9 2.7 4.4	In-Ceram® Alumina In-Ceram® Spinell In-Ceram® Zirconia	Vident Vident Vident
Metal-ceramic	Sintering	Leucite ($KalSi_2O_6$) Leucite ($KalSi_2O_6$)	61 ± 5 61 ± 5	15–25 15–25		VMK-95 Ceramco®3	Vident Dentsply

ceramic bridges, and metal-ceramic prostheses. The microstructure of ultralow-fusing ceramics demonstrates small crystal particles thoroughly distributed in the structure or even few or no crystals and are more polishable than low-fusing and medium-fusing ceramics. These ceramics generally contain high concentrations of CaO, K_2O, Li_2O, and Na_2O. Moreover, due to the reduced amounts of leucite phase present in the composition of these ceramics, they exhibit lower expansion and contraction coefficients [1, 8].

3.1.3.3 Based on Fabrication Method

The most practical way of classifying dental ceramics is based on the fabrication method. The fabrication process of ceramics is a critical factor affecting the microstructure, volume, and quantity of formed cracks, voids, and inclusions, which could accelerate crack propagation and failure of the fabricated dental ceramic. Dental ceramics can be fabricated using various methods including sintering, glass infiltration, slip casting, heat pressing, CAD-CAM machining, copy-milling, and combined techniques. Ceramics fabricated in the same method usually show similar behaviors of polishability, failure mechanisms, esthetics, wear resistance, and abrasiveness [1, 3, 9].

3.1.3.4 Based on Microstructure and Composition

Mechanical and optical properties of dental ceramics are associated with composition, microstructure, and defects presented in the structure. All dental ceramics apart from the fabrication method could exhibit four main microstructures, which are (I) glass-based systems (ex. Silica), (II) glass-based systems with fillers typically crystalline (ex. Leucite and lithium disilicate), (III) crystalline-based systems with glass fillers (ex. Alumina), and (IV) polycrystalline solids (ex. Alumina and zirconia) [9–11].

Composition Category 1. These glass-based systems are usually obtained from feldspars and contain silica as the main constituent besides alumina, potassium, and sodium with low thermal expansion coefficients (8×10^{-6} 1/K). Predominantly glass materials mainly composed of amorphous glass matrix, and have the lowest mechanical properties compared to other ceramics. However, from the optical standpoint, these dental ceramics typically match better to enamel and dentin appearance. These ceramics are generally used for veneering alumina-based cores [9, 10].

Composition Category 2. These glass-based systems contain different types of fillers, and silica is the main component of glass. The crystalline phase present in the microstructure of dental ceramics acts as a reinforcement and provides strength and toughness; however, it can lower its optical properties by making it opaque. Precipitation of these particles among the base glassy matrix with the help of special heat treatments leads to the formation of glass-ceramics. In general, the glass phase brings about brittleness and translucency to the ceramic while the crystalline phase presents opacity, color shade, chemical stability, and mechanical strength to the material. Fillers are usually crystalline leucite, lithium disilicate, or fluoroapatite with varying crystal to glass ratios creating three different subcategories, which are low-to-moderate leucite-containing feldspathic glass (feldspathic porcelains), high leucite-containing glass-ceramic (about 50%), and lithium-disilicate glass-ceramic. The glass composition of these subcategories is usually similar to that of pure glass of the previous category [9–11].

Composition Category 3. This category comprises of crystalline-based systems with glass fillers. The structure of these materials is composed of a sintered crystalline matrix surrounding particles in the crystalline phase. Glass-infiltrated partially sintered alumina (In-Ceram), which was introduced in 1988 is a member of this composition category produced from a porous alumina core infiltrated with low-viscosity lanthanum oxide (La_2O_3) glass at high temperature. The crystalline phase of the framework can be alumina, alumina/zirconia, or alumina/magnesia spinel fabricated by slip casting or milling of a pre-sintered block. The interpenetration of glass into the porous structure of the core improves its fracture toughness (2.48–3.55 $MPa.m^{1/2}$) and flexure strength (300–500 MPa) [9, 11].

Composition Category 4. Polycrystalline solids of alumina and zirconia are classified in this category. These materials are solid-sintered, monophase ceramics formed by direct sintering of crystals using high sintering temperatures. These glass-free polycrystalline materials have great strength and toughness. However, the necessity of high temperatures has prohibited the use of these high-strength frameworks for crowns and FPDs [9]. Table 3.3 lists different applications, fabrication techniques, and crystalline phases of dental ceramics.

3.1.4 Metal-Ceramics (MCs)

Brecker in the 1950s introduced metal-ceramic dental restorations by a process of baking porcelain onto gold. These kinds of restorative materials are composed of a metal alloy substructure (coping) such as Co–Cr or Au–Pd produced by the casting method and several veneering ceramic layers, which are fused onto the metal coping to cover the color of the metal substrate and provide good esthetic. Opaquer is the first layer applied to the metal framework to cover its metallic color as well as providing a bond between the metal framework and veneering porcelain. Dentine and enamel porcelain with different levels of translucency are the next layers applied on the opaque layer [8, 13]. Metal-ceramic (MC) prostheses have shown great survivability through clinical studies because of their outstanding properties. Contrary to the all-ceramic restorations, these restorative materials are ductile and application of them requires less tooth removal. That is why porcelain-fused-to-metals (PFM) materials or metal-ceramic (MC) systems with combined properties of the good strength and toughness resulted from the metal substructure and appropriate esthetics of the porcelain veneer are popular. These restorations present good marginal fit and are easy to repair their chipped or delaminated veneer. Moreover, pressing the ceramic to the framework is rapid and saves time. Metal-ceramic restorations are suitable for long-span frameworks (5 and more). Ceramics used to cover the inappropriate esthetics of the metal bases are composed of silica, alumina, and potassium-oxide. Leucite, lithia disilicate, alumina, zirconia, and apatite are the possible crystal phases of these veneering ceramics. Potassium feldspar and leucite are the main products of firing the veneer precursors on the metalcore. Esthetics of metal-ceramic restorations is not as pleasant as all-ceramic restorations presenting inappropriate translucency, grayish color, and dark line at the facial margin of the metal-ceramic crown associated with a metal collar or metal margin. Besides this, some patients have reported metal allergy situations due to the nickel content of the metal substructure. The common failure of the metal-ceramic restorations is debonding of the veneering ceramic from the metal substructure due to the different materials with different coefficients of thermal expansion. On the other hand, casting frameworks are a time-consuming process and the metal frameworks are usually expensive [1, 13, 14].

Feldspathic Veneers of MCs. Feldspathic porcelains are the most traditional ceramics based on the leucite phase used for veneering metal-ceramic crowns and fixed dental prostheses (FDP) from 1970s. The term "porcelain" is generally used for ceramics composed of potash or soda feldspars ($K(Na)AlSi_3O_8$ or $K(Na)_2O.Al_2O_3.6SiO_2$) as the base mineral accompanied with lower amounts of kaolinite clay ($Al_2Si_2O_5(OH)_4$ or $Al_2O_3.2SiO_2.2H_2O$); and quartz (SiO_2), which are mixed and fired at high temperatures. Chemical composition of feldspathic porcelains includes SiO_2 (52–65 wt%), Al_2O_3 (11–20 wt%), K_2O (10–15 wt%), and Na_2O (4–15 wt%). The composition of feldspathic veneers of metal-ceramic restorations includes more quantities of K_2O and Na_2O compared to those of other ceramic veneers in order to decrease the sintering temperature and increase the thermal expansion coefficient of the (8×10^{-6} K^{-1}) ceramic to better match with the metalcore. Some amounts of additives such as pigments are also used in the formulation of glass to provide desired color shades and cover the underlying metallic color. Oxide of metals such as iron, nickel, copper, titanium, manganese, and cobalt are some of the examples used in the composition of porcelains. Cerium oxide, zirconium oxide, titanium oxide, and tin oxide are glass opacifier additives and introduced the first feldspathic crowns followed by increasing the interest and demand for ceramic restorative materials. These feldspar-based ceramics contain a minimum of 15 wt% silica as reinforcement providing translucency and low-shrinkage behavior to the product as well as less than 4 wt% kaolin to provide green strength of the product. In feldspathic veneering porcelains, alkali cations (sodium or potassium) fill the sites among the tetrahedrons to balance the electrical charge of the whole structure. These cations could break silicate chains leading to thermal expansion increase of silicate glass. Figure 3.5 illustrates the 2D structure of a potassium silicate glass [1, 8, 9, 15].

Firing potash feldspar at 1150 °C leads to the formation of leucite ($KalSi_2O_6$) crystals. Feldspathic dental porcelains usually contain between 15 and 25 vol.%

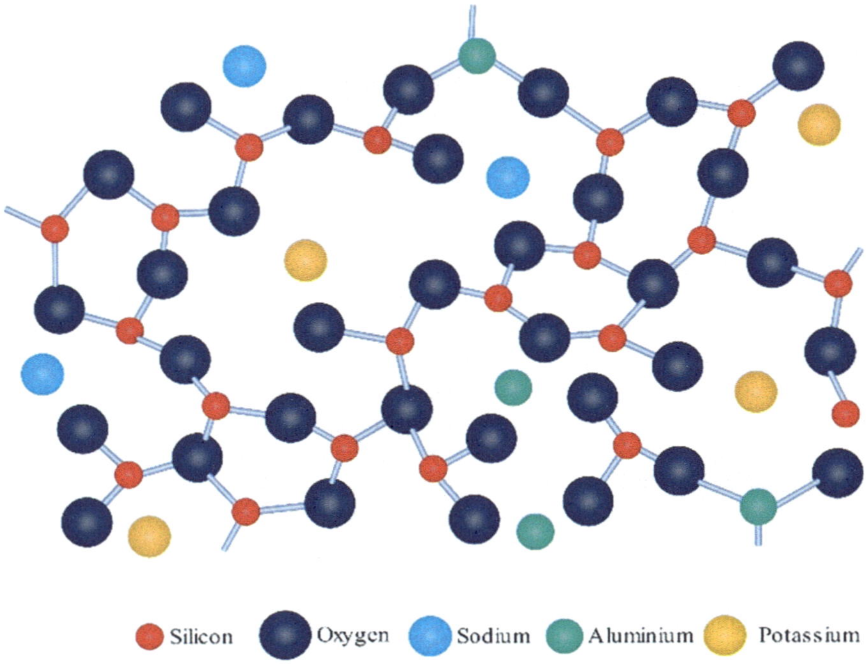

Fig. 3.5 2D structure of amorphous potassium silicate glass. Redrawn from Kelly et al. [16]

leucite [8]. A displacive phase transformation from tetragonal to cubic at 625 °C, accompanied by volume expansion of 1.2% occurs in leucite structure. Crystalline particles of leucite have the ability to increase the thermal expansion coefficient and enhance the esthetics of porcelains. By modifying the proportion of leucite to feldspar glass, the coefficient of thermal expansion can be precisely adjusted to match the metal substructure and reduce the risk of crack formation. The values of the coefficient of thermal expansion for feldspathic porcelains are about 20–25 × 10^{-6} °C^{-1}. Nowadays, most of the veneering porcelains are obtained from potash feldspar (sanidine), which is stable at high temperatures [12]. The high translucency, fluorescency, opalescency, and thermal expansion coefficient, as well as low density, chemical stability, longevity, and biocompatibility of leucite-contained feldspathic porcelains, facilitates their use with metals, alumina, and zirconia, which was a starting point in future developments of dental ceramics. However, due to the large particle size of leucite crystals, these materials have low toughness, low strength, and cause enamel abrasion. Besides these, low marginal accuracy and great risk of surface microcracks are other drawbacks of these porcelains, which limit their use in high-bearing posterior regions and that is why these porcelains are more used as veneers on minimally prepared anterior teeth. Because of the large difference between the coefficient of thermal expansion of leucite crystals and the glassy matrix, radial tensile stresses and tangential compressive stresses mode within and around the crystals while cooling, which results in the formation of microcracks. The mechanical properties of feldspathic porcelains are the lowest compared to the other ceramic restoratives. Keeping in mind that metal-ceramic materials have been successfully used in restorative dentistry for more than four decades, their overall performance has been approved [1, 3, 15].

3.1.5 All-Ceramics

Dental ceramics have been widely used and evaluated through the past forty years in dental applications due to their outstanding clinical, physical, optical, and mechanical properties. The use of ceramics in dentistry has been rapidly moved forward from the introduction of traditional feldspathic porcelains to the application of modern ceramics with enhanced strength, toughness, and esthetics. The development of new-generation ceramics for dentistry has been always a challenging topic requiring a careful and time-consuming study on ceramic composition, microstructure, crystalline phase content, mechanical properties, biological response, and processing method to fulfill the required demands of patients, and clinical technicians. As, the hardness lower than that of tooth enamel, convenient esthetics and chemical stability should be manipulated in a less invasive, less painful, and more rapid manner [1, 17]. Nowadays utilization of traditional metal-ceramic materials is being replaced by high-performance all-ceramic restorations with high strength, toughness, and esthetics. The development of metal-free restorations in dentistry was a consequence of the increasing demand for utilizing restorations with better mechanical and optical properties over the last three decades. All-ceramic dental restorations present a high color match to the natural human teeth and outstanding mechanical properties. Contrary to the metal-ceramic materials, all-ceramic materials consist of a large amount of crystalline phase (35–99 vol.%) providing high mechanical properties. However, the amount of the crystalline phase present within the glass matrix has a direct influence on the opacity of dental ceramics [1]. Alumina-reinforced leucite-based ceramic was the first product of all-ceramic restorations. Glass-infiltrated ceramics including VITA In-ceram Alumina, Spinell, and Zirconia (VITA Zahnfabrik) with higher strength and optical properties than feldspathic porcelains were introduced afterward. Nowadays, these restorative materials are used as inlays, onlays, veneers, and crowns. All-ceramic crowns offer greater potential for success in matching the appearance of the adjacent natural tooth, especially when a relatively high degree of translucency is desired. All-ceramic restorations composed of veneering porcelain with high esthetics accompanied with ceramic cores demonstrating excellent flexural strength have been recently developed. The crystalline phase of the all-ceramics materials can be lithium disilicate glass-ceramics, alumina/zirconia, and zirconia. Aluminum and zirconium oxides as the ceramic materials are popular in dental materials, which can be produced using various procedures including sintering, slip casting, heat pressing, and CAD/CAM technique. The fabrication of dental implant abutments and FDPs have been developed by introducing high-strength zirconia-based systems. In addition, ceramics are still used to fabricate denture teeth. All-ceramic systems are categorized into six different classes including HIP or CAD/CAM core ceramics, manually condensed veneering ceramics, hot-pressed veneering ceramics, liners, ceramic stains, and ceramic glazes. Application of the CAD/CAM technique besides the use of monolithic ceramics with high fracture toughness is the latest trend in restorative dentistry. The common clinical complication associated with the use of all-ceramic restorations is the fracture of veneering porcelain or the substructure. Moreover, for aluminum oxide and lithium disilicate FPDPs, the fracture of the connector is the major reason of failure; and for the zirconia FPDPs, the fracture of veneering porcelain is the main reason of failure. However, in the metal-ceramic FPDPs, the main reason of failure is tooth fracture and caries. The success of the use of these materials depends on the procedures utilized to prevent crack propagation through fabrication process as well as the capability of clinicians to properly select the materials, cements, procedures, and the thickness ratio of the ceramic core to the veneer [12, 14, 15]. Table 3.4 lists different all-ceramic core materials besides their manufacturing technique and recommended clinical indications.

3.1.5.1 Glass-Ceramics

Glass and glass-ceramics are important materials used in restorative dentistry, thanks to their biocompatibility, long-term durability, excellent esthetics, good mechanical strength, and relative ease of use. High portions of the crystalline phase within the glassy matrix of these materials increase their toughness to a large extent and hinder crack growth. MacCulloch, in 1968 introduced and used glass-ceramics for posterior denture teeth. Later in the 1970s, McLean and O'Brian developed the use of glass-ceramics by introducing leucite-based glass-ceramics used as veneers. By extra studies on the basis of producing polycrystalline ceramics, Sadoun in 1987 introduced glass-infiltrated alumina cores with the name of In-Ceram fabricated by slip-casting technique. Later, in 1993 Andersson and Odén introduced Procera® (Nobel Biocare, Kloten, Switzerland) made of porcelain veneered polycrystalline alumina with high purity. Dicor (Dentsply International, York, PA, USA), a micaceous glass-ceramic was the first commercial product of glass-ceramics, which was produced using lost-wax and centrifugal casting methods composed of 45 vol.% glass and 55 vol.% tetrasilicic fluormica. Dicor presents suitable flexural strength (120–150 MPa) applied as crowns, veneers, and inlays in the anterior regions owing to its good esthetics. However, the use of Dicor ceramic obsoleted soon after its introduction due to its low tensile strength, low fracture toughness, and time-consuming complex clinical application compared to those of PFM restorations. Modern glass-ceramics developed in the 1990s by introducing heat-pressed ceramics with easier fabrication procedures and better overall properties. Leucite-reinforced (IPS Empress®; Ivoclar, Schaan, Liechtenstein) and lithium disilicate-reinforced (IPS Empress® II, IPS e.max®; Ivoclar, Schaan, Liechtenstein) glass-ceramics are examples of these pressable ceramics providing higher fracture toughness. The values of fracture

Table 3.4 Ceramic materials and systems and manufacturer-recommended clinical indications (Reprinted from The Journal of Prosthetic Dentistry, Conrad et al., Current ceramic materials and systems with clinical recommendations: a systematic review, no. 390, Copyright (2007), with permission from Elsevier) [15]

Core material	System	Manufacturing technique	Clinical indications
Glass-ceramic			
Lithium-disilicate (SiO_2–Li_2O)	IPS Empress 2 (Ivoclar Vivadent, Schaan, Liechtenstein)	Heat pressing	Crowns, anterior FPDP
	IPS e.max Press (Ivoclar Vivadent)	Heat pressing	Onlays, 3/4 crowns, crowns, FPDP
Leucite (SiO_2–Al_2O_3–K_2O)	IPS Empress (Ivoclar Vivadent)	Heat pressing	Onlays, 3/4 crowns, crowns
	Optimal Pressable Ceramic (Jeneric Pentron, Wallingford, Conn)	Heat pressing	Onlays, 3/4 crowns, crowns
	IPS ProCAD (Ivoclar Vivadent)	Milling	Onlays, 3/4 crowns, crowns
Feldspathic (SiO_2–Al_2O_3–Na_2O–K_2O)	VITABLOCS Mark II (VITA Zahnfabrik, Bad Sackingen, Germany)	Milling	Onlays, 3/4 crowns, crowns, veneers
	VITA TriLuxe Bloc (VITA Zahnfabrik)	Milling	Onlays, 3/4 crowns, crowns, veneers
	VITABLOCS Esthetic Line (VITA Zahnfabrik)	Milling	Anterior crowns, veneers
Alumina			
Aluminum oxide (Al_2O_3)	In-Ceram Alumina (VITA Zahnfabrik)	Slip-casting, milling	Crowns, FPDP
	In-Ceram Spinell (VITA Zahnfabrik)	Milling	Crowns
	Synthoceram (CICERO Dental Systems, Hoorn, The Netherlands)	Milling	Onlays, 3/4 crowns, crowns
	In-Ceram Zirconia (VITA Zahnfabrik)	Slip-casting, milling	Crowns, posterior FPDP
	Procera (Nobel Biocare AB, Goteborg, Sweden)	Dense sintering	Veneers, crowns, anterior FPDP

(continued)

Table 3.4 (continued)

Core material	System	Manufacturing technique	Clinical indications
Zirconia			
Yttrium tetragonal zirconia polycrystals (ZrO_2 stabilized by Y_2O_3)	Lava (3 M ESPE, St. Paul, Minn)	Green milling, sintering	Crowns, FPDP
	Cercon (Dentsply Ceramco, York Pa)	Green milling, sintering	Crowns, FPDP
	DC-Zirkon (DCS Dental AG, Allschwil, Switzerland)	Milling	Crowns, FPDP
	Denzir (Decim AB, Skelleftea, Sweden)	Milling	Onlays, 3/4 crowns, crowns
	Procera (Nobel Biocare AB)	Dense sintering, milling	Crowns, FPDP, implant abutments

toughness for these two ceramics range between 1–6 $MPa.m^{1.2}$ and 1.8 $MPa.m^{1.2}$ for leucite-reinforced and 2.8–3.5 $MPa.m^{1/2}$ for lithium disilicate-reinforced glass-ceramics. The high amount of fracture toughness for the latter glass-ceramic is due to the larger amounts of crystalline phase within the lithium disilicate glass-ceramic and formation of interlocked plate-like crystals. Moderate strength of IPS Empress ceramic (120–180 MPa) limits their use in single unit complete-coverage restorations in the anterior regions [1, 13, 14].

Feldspathic Glass-Ceramics. Vita Mark II (VITA Zahnfabrik, Bad Sackingen, Germany) is an important feldspathic glass-ceramic with high machinability characteristic due to its fine grain size which is about 4 μm and is suitable to be used with the CEREC® CAD/CAM system (Sirona Dental Systems, Charlotte, NC) for fabricating inlays and onlays comparing strongly with metal-ceramics. This glass-ceramic can be produced in different color shades such as Classic Line Vita shades, Vitapan 3Dmaster Shades, VITABLOCS Esthetic Line, and a bleached shade. Figure 3.6 illustrates the microstructure of a feldspar-based Vitablocs Mark II (Vident) consisting of fine polygonal sanidine crystals spread within a glass phase [9, 15].

Fluormica Glass-Ceramics. Grossman in 1972 introduced the first dental glass-ceramic based on tetrasilicic fluormica with the commercial name of Dicor® for application of crowns, veneers, inlays, and onlays of anterior regions. Dicor® glass-ceramic restorations can be produced using either casting of the glass ingots or CAD/CAM machining. This glass-ceramic includes 55 vol.% tetrasilicic fluormica platelets ($K_2Mg_5Si_8O_{20}F_4$). The microstructure of a fluormica glass-ceramic is shown in Fig. 3.7. Most of the evolution of this glass-ceramic is associated with its high translucency, excellent chemical resistance, appropriate machinability, and acceptable flexural strength (150 MPa). Fine crystal size (about 1 μm) of this type of material brings about excellent translucency as well as high machinability and strength to the micaceous glass-ceramic [7].

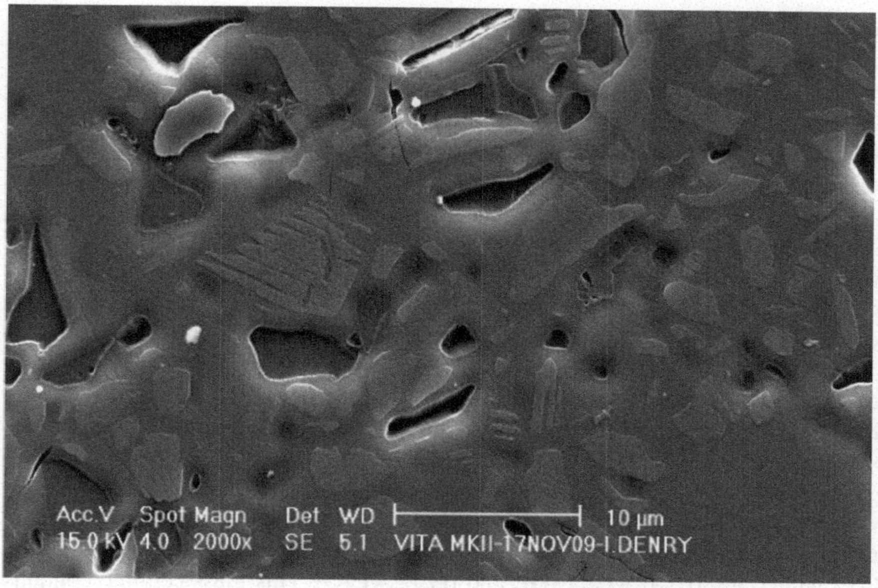

Fig. 3.6 The microstructure of a fully sintered machinable feldspar-based ceramic (Vitabloc Mark II). (Reprinted from Materials, Denry, & Holloway, Ceramics for dental applications: a review, no. 359, Copyright (2010), with permission from MDPI, CC BY 3.0) [12]

Leucite-Containing Glass-Ceramics. Leucite-based ceramics are an example of glass-ceramics manufactured by controlled nucleation and crystallization of crystalline phase in the glassy matrix. The microstructure of a leucite-containing glass-ceramic is composed of an aluminosilicate glass matrix accompanied by about 50% of leucite crystalline phase (for high leucite-containing glass-ceramics) grown within the glass via controlled crystallization method. IPS Empress (Ivoclar Vivadent, Amherst, NY) is the most important example of these glass-ceramics. These materials could be fabricated as pressable or machinable glass-ceramics highly employable with CEREC and E4D (D4D Technologies, LLC, Richardson, TX) for fabrication of posterior inlays and onlays as well as anterior veneers and crowns. The utilized fabrication technique has the most effect on the fracture resistance and strength of low, moderate, and high-containing glass-ceramics rather than the crystal type, amount, and distribution. For example, machinable and pressable systems show higher fracture resistance than powder/liquid systems with better clinical results when used as posterior inlays, onlays, anterior veneers, and crowns. Figure 3.8 shows the microstructure of a heat-pressed leucite-reinforced glass-ceramic [9].

The high strength of leucite-reinforced glass-ceramics (approximately 180 MPa in biaxial tests) is a result of the stress formation around the glass matrix as a consequence of the higher CTE of leucite crystals compared to those of glass. Besides this, the appropriate translucency and wear resistance of this kind of glass-ceramics have made them a good choice for metal-free dental restorations of inlays, onlays, and crowns especially in the anterior areas [2]. IPS ProCAD (Ivoclar Vivadent), a leucite-reinforced ceramic similar to IPS Empress, has fine particle size and was introduced in 1998 to be used with the CEREC inLab system (Sirona Dental Systems, Bensheim, Germany). IPS ProCAD is available in numerous shades [15]. IPS InLine® POM (Ivoclar Vivadent AG, Liechtenstein) is a leucite-based glass-ceramic suitable to be used via the press on metal method, which provides fusing of the dense materials on a metal substrate as well as more rapid and precise design and manufacturing process. The chemical composition of this veneering ceramic includes 50–60 wt% SiO_2, 8–20 wt% Al_2O_3, 7–13 wt% K_2O, and 4–12 wt% Na_2O. This glass-ceramic, which is used for layering and sintering upon metals presents a high thermal expansion coefficient (equal to $12.6 \pm 0.5 \times 10^{-6}\ K^{-1}$ matching to that of metal substrates, which is about 13.8×10^{-6}–16.2×10^{-6} 1/K), good flexural strength (80 MPa), and good chemical durability bonded to the metal through an intermediate layer, which also covers the metal. IPS InLine can be used to veneer metal crowns and long-span bridges of the posterior regions. Contrary to the IPS InLine glass-ceramic, the IPS d.SIGN glass-ceramic is suitable to be used for veneering metal frameworks in the anterior regions and long-span bridges. The chemical composition of this veneering ceramic includes 49–58 wt% SiO_2, 11–21 wt% Al_2O_3, 9–23 wt% K_2O, 1–10 wt% Na_2O, 2–12 wt% CaO, 0.5–6 wt% P_2O_5, and 0.2–2 wt% F. This glass-ceramic contains leucite and fluoroapatite ($Ca_5(PO_4)_3F$) crystals after firing (Fig. 3.9) and exhibits

Fig. 3.7 Image of the microstructure of machinable tetrasilicic fluormica glass-ceramic (Dicor® MGC-light, magnification × 2000 and field of view: 60 μm × 60 μm). (Reprinted from Butterworth-Heinemann, Saint-Jean, S. J., Dental glasses and glass-ceramics, In Advanced Ceramics for Dentistry, no. 263, Copyright (2014), with permission from Elsevier) [7]

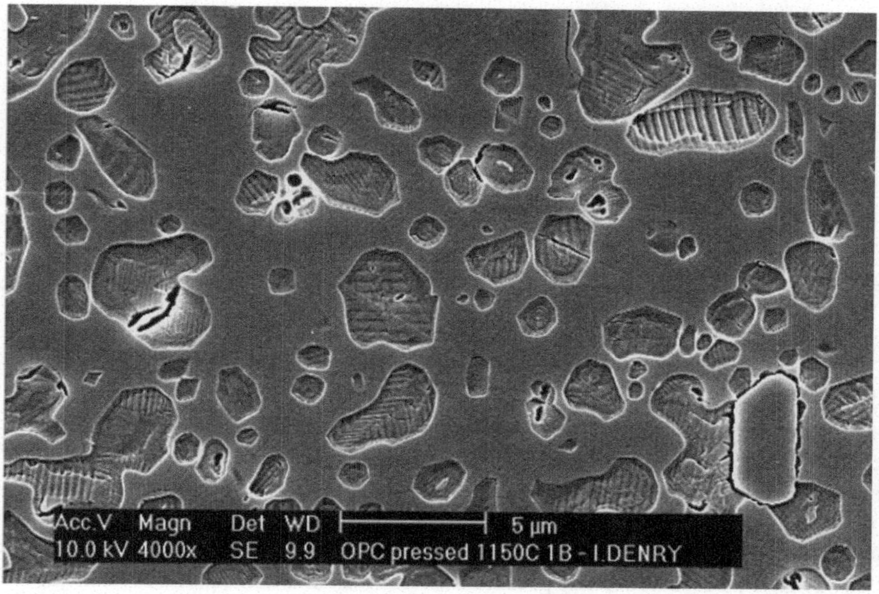

Fig. 3.8 Microstructure of heat-pressed leucite-reinforced glass-ceramic of IPS Empress. (Reprinted from Materials, Denry, I., & Holloway, J. A., Ceramics for dental applications: a review, no. 354, Copyright (2010), with permission from MDPI, CC BY 3.0) [12]

good flexural strength (about 80 MPa), good chemical stability, convenient coefficient of thermal expansion ($13.0 \pm 0.5 \times 10^{-6}$ K^{-1}), and excellent tooth-similar color and translucency (Fig. 3.10) [2, 14].

Lithium-Disilicate Glass-Ceramics. Glass-ceramics have gained much more attraction after the introduction of lithium disilicate glass-ceramics in 1998 (IPS Empress® 2, Ivoclar Vivadent Ltda, Schaan, Liechtenstein, later on marketed as e.max®) fabricated using a combination of lost-wax and heat-pressed techniques by adding lithium oxide to the aluminosilicate glass. This lithium-disilicate glass-ceramic was based on the system of $Li_2O{:}2SiO_2$. The raw powder of lithium-disilicate glass-ceramic (IPS Empress® 2) is consisting of SiO_2 (57–80 wt%), Al_2O_3 (0–5 wt%), La_2O_3 (0.1–6 wt%), MgO (0–5 wt%), ZnO (0–8 wt%), K_2O (0–13 wt%), Li_2O (11–19 wt%), P_2O_5 (0–11 wt%), and additives (about 8 wt%). Figure 3.11 shows the microstructure of heat-pressed IPS Empress 2, which is consisting of interlocked needle-like crystals. IPS Empress® 2 glass-ceramics are successfully used as crown and bridge frameworks of the anterior regions [14, 17].

Lithium disilicate-based glasses have superior chemical stability, flexural strength (550 MPa according to dental standard ISO 6872), and toughness (K_{Ic} = 2.3–2.9 MPa.$m^{1.2}$ according to dental standard ISO 6872) compared to leucite glass-ceramics [2]. The values of flexural strength and fracture toughness for leucite glass-ceramic are about 164 MPa and 1.03 MPa.$m^{1.2,}$ respectively, while these values vary to about 365 MPa and 2.80 MPa.m$^{1.2}$, respectively, for lithium disilicate glass-ceramics [17].

IPS e.max products can be produced in both pressable and machinable forms. IPS e.max Press (Ivoclar Vivadent) was introduced in 2005 as an improved press-ceramic material compared to IPS Empress 2 consisting of a lithium-disilicate pressed glass-ceramic with modified physical and optical properties. IPS e.max CAD is the other product of the Ivolcar Vivadent Company with enhanced machinability properties. The high mechanical properties of the IPS e.max CAD lithium-disilicate glass-ceramic is associated with the presence of high amounts (nearly 60%) of needle-like silicate crystals and interlocked microstructure leading to a great energy loss of cracks during their path around the crystals (Fig. 3.12). The flexural strength and fracture toughness of this kind of glass-ceramic are equal to 440 MPa and 2.3–2.9 MPa.m$^{1/2}$, respectively [14, 15]. IPS e.max products are highly translucent. The high translucency of this material owes to low refractive index of the lithium-disilicate crystals even with high content of the crystalline phase and can be employed for full-contour restorations [9].

Aesthetics of lithium disilicate is not as perfect as feldspar-based ceramics; however, in case of high esthetics demands it can be modified using fluorapatite [$Ca_5(PO_4)_3F$]-based veneering porcelain with low contents of apatite crystals joined to lithium-disilicate core by sintering or molding process (Fig. 3.13). These glass-ceramics are useful for short-span 3-unit fixed partial dentures in the anterior regions,

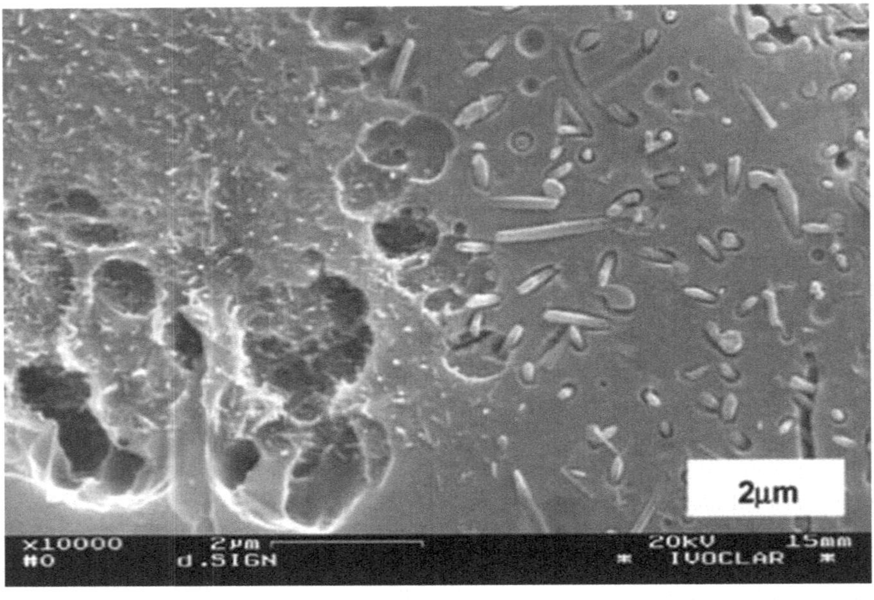

Fig. 3.9 Microstructure of a leucite-apatite glass-ceramic. (Reprinted from Journal of Materials Science: Materials in Medicine, Holand et al., Clinical applications of glass-ceramics in dentistry, no. 1040, Copyright (2006), with permission from Springer) [18]

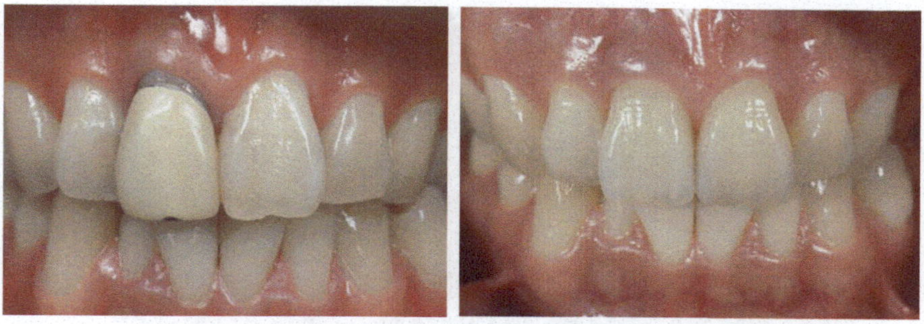

Fig. 3.10 View of a tooth requiring restoration before (left) and after application of leucite-fluoroapatite glass-ceramic IPS d-SIGN® as a veneer on metal (right) restoration. (Reprinted from Journal of Materials Science: Materials in Medicine, Holand et al., Clinical applications of glass-ceramics in dentistry, no. 1040, Copyright (2006), with permission from Springer) [18]

Fig. 3.11 Microstructure of heat-pressed lithium-disilicate glass-ceramic (IPS empress 2). (Reprinted from Dental Clinics of North America, Zhang & Kelly, Dental ceramics for restoration and metal veneering, no. 804, Copyright (2017), with permission from Elsevier) [5]

posterior crowns, and all-ceramic bridges [2]. The typical composition of fluorapatite glass-ceramic consists of 60.0–65.0 wt% SiO_2, 8.0–12.0 wt% Al_2O_3, 6.0–9.0 wt% Na_2O, 6.0–8.0 wt% K_2O, 2.0–3.0 wt% ZnO; and some amounts of CaO, P_2O_5, F, other oxides, and pigments [7]. Figure 3.14 shows the microstructure of fluoroapatite glass-ceramics.

Although lithium-disilicate glass-ceramics are widely used with great success, some efforts have been performed to minimize the remaining disadvantages of

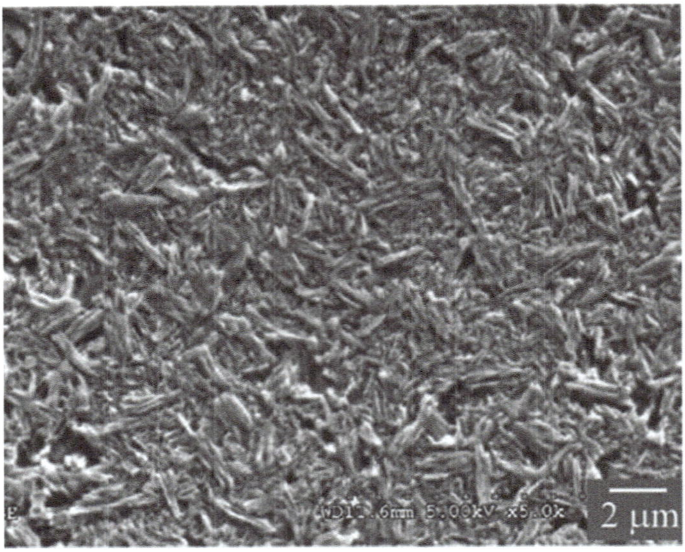

Fig. 3.12 Microstructure of lithium-disilicate glass-ceramic (IPS e.max CAD). (Reprinted from Dental Clinics of North America, Zhang & Kelly, Dental ceramics for restoration and metal veneering, no. 804, Copyright (2017), with permission from Elsevier) [5]

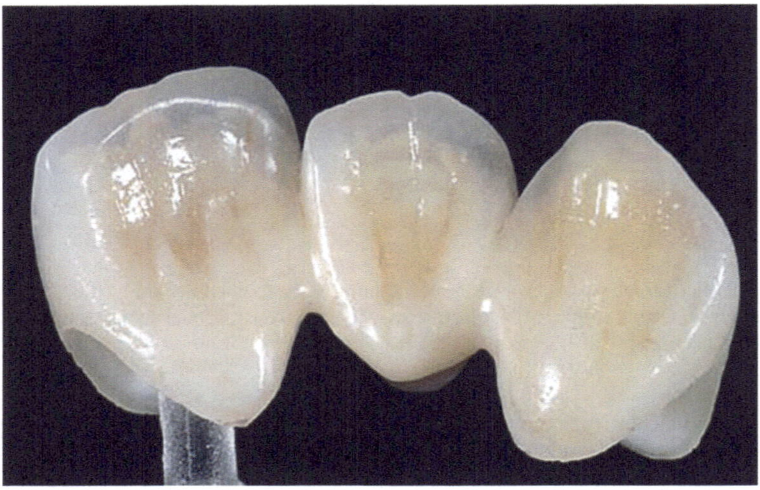

Fig. 3.13 Three-unit dental bridge of lithium-disilicate glass-ceramic framework (IPS empress®II) veneered with apatite-containing glass-ceramic (IPS ERIS®). (Reprinted from Journal of Materials Science: Materials in Medicine, Holand et al., Clinical applications of glass-ceramics in dentistry, no. 1039, Copyright (2006), with permission from Springer) [18]

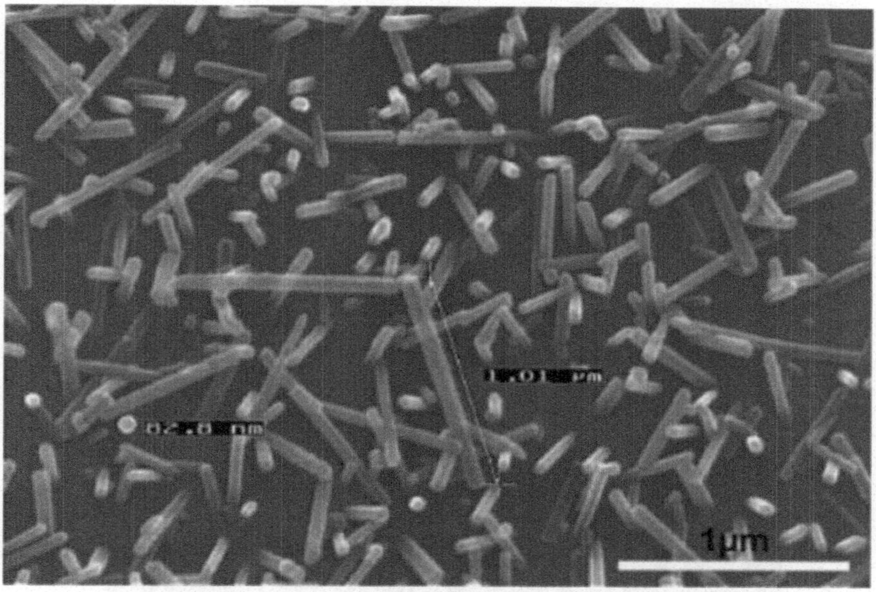

Fig. 3.14 The microstructure of fluorapatite glass-ceramic used for veneering high-strength frameworks. (Reprinted from Butterworth-Heinemann, Saint-Jean, S. J., Dental glasses and glass-ceramics, In Advanced Ceramics for Dentistry, no. 269, Copyright (2014), with permission from Elsevier) [7]

these ceramic systems with the help of polycrystalline-reinforced zirconia-containing lithium-silicate ceramics (ZLS). ZlS is a novel machinable material exhibiting good optical properties and superior mechanical properties and is easily milled using CAD-CAM machines. ZLS is based on lithium-silicate glass with 10 wt% zirconia addition as a nucleation agent. Suprinity (Vita Zahnfabrik, Bad Sachingen, Germany) and CELTRA Duo (Dentisply-Sirona, Bensheim, Germany) are examples of these glass-ceramics applied in restorative dentistry accompanied with the CAD-CAM techniques due to their ease of milling. The microstructure of this kind of glass-ceramics contains fine and round-shaped lithium metasilicate (Li_2SiO_3) crystals, rod-like lithium disilicate ($Li_2Si_2O_5$) crystals, and zirconia-containing glass matrix, which is shown in Fig. 3.15 [19].

While lithium disilicate glass-ceramics (L2S) contain only lithium disilicate crystals, the machinable ZLS glass-ceramics contain lithium metasilicate crystals in addition to lithium disilicate crystals as a result of second crystallization step of heat treatment at 840 °C for 8 min. Through the crystallization process of ZLS, formation of smaller lithium-silicate crystals with size ranges of 0.5–1 μm occurs owing to the presence of zirconia particles. The main advantage of these materials is their timesaving ability for the production of dental restorations, and are already offered in their fully crystallized state (CELTRA Duo, Dentisply-Sirona, Bensheim, Germany), no furnace need state or the state requiring a very short crystallization

Fig. 3.15 The microstructure of zirconia lithium-silicate machinable glass-ceramic Suprinity (Vita). Reprinted from Dental Materials, de Carvalho Ramos et al., Microstructure characterization and SCG of newly engineered dental ceramics, no. 875, Copyright (2016), with permission from Elsevier) [20]

cycle (Suprinity, Bad Sachingen, Germany) [19]. The composition of these materials are similar and are commonly consisted of SiO_2 (56–64 wt%), Al_2O_3 (1–4 wt%), CeO_2 (0–4 wt%), ZrO_2 (8–12 wt%), K_2O (1–4 wt%), Li_2O (15–21 wt%), and P_2O_5 (3–8 wt%) [17].

3.1.5.2 Glass-Infiltrated Ceramics

Glass-infiltrated ceramics are crystalline-based systems composed of a sintered porous core (framework) infiltrated with a low-viscosity lanthanum glass at high temperatures. The crystalline phase of the framework can be alumina, alumina/zirconia, or alumina/magnesia spinel fabricated by slip casting or milling of a pre-sintered block. The special structure of these ceramics provides improved fracture toughness (2.48–3.55 $MPa.m^{1/2}$) and flexure strength (300–500 MPa) due to the crystalline base of structure with low glass phase content. These ceramics present convenient fit of the prostheses owing to the low sintering shrinkage [1, 8, 9].

In-Ceram Alumina containing about 85 vol.% alumina was the first glass-infiltrated partially sintered ceramic introduced in 1988 for restoring single and three-unit fixed partial dentures of anterior and posterior crowns. The ceramic core can

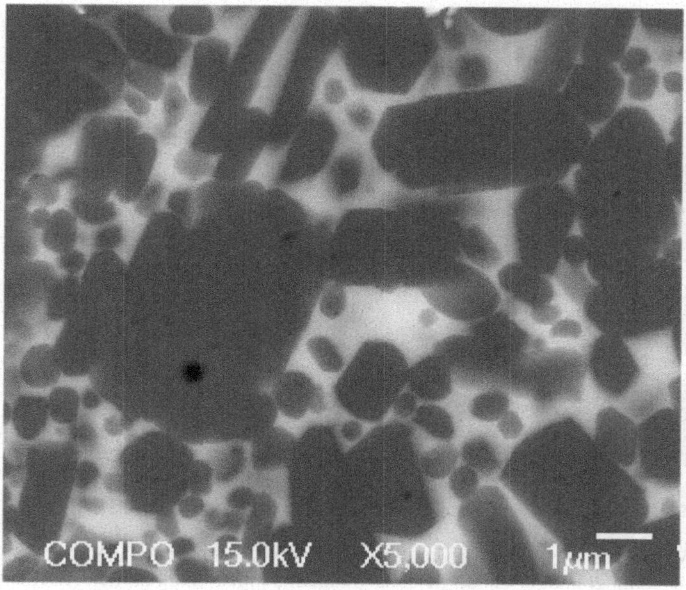

Fig. 3.16 Microstructure of glass-infiltrated Alumina (In-Ceram Alumina) illustrating crystalline-based structure associated with glass filler. (Reprinted from Butterworth-Heinemann, Saint-Jean, S. J., Dental glasses and glass-ceramics, In Advanced Ceramics for Dentistry, no. 269, Copyright (2014), with permission from Elsevier) [7]

be either fabricated using the slip-casting technique or copy-milling from partially sintered blanks [9]. Through the slip-casting process of In-Ceram Alumina, a porous partially sintered skeleton of alumina is fabricated by the application of a densely packed alumina slurry with 70–80 wt% to a refractory die followed by its sintering at 1120 °C for 10 h. The lanthanum glass is then infiltrated into the porous alumina at 1100 °C for 4–6 h leading to the elimination of formed porosities, increasing the strength, and decreasing the risk of crack propagation (Fig. 3.16) [15]. As the sintering temperature of In-Ceram crowns is relatively low, this material undergoes little shrinkage during sintering resulting in the production of ceramics with favorable dimensions and superior marginal fit. The formed coping is then veneered with feldspathic porcelain such as Vitadur Alpha (Vita Zahnfa-brik) [1]. The flexural strength of In-Ceram Alumina framework varies from 236 to 600 MPa. Furthermore, its fracture toughness is about 3.1–4.61 MPa.m$^{1.2}$ [21].

In-Ceram Zirconia (VITA Zahnfabrik) was the first all-ceramic product used for the fabrication of single-unit restorations and 3-unit FPDs of anterior and posterior regions consisting of glass-infiltrated sintered alumina with 35 wt% of partially stabilized zirconia (Fig. 3.17). Similar to In-Ceram Alumina, the core of these products can be fabricated using the slip-casting technique and copy-milling from partially sintered blanks. Flexural strength and fracture toughness of the In-Ceram Zirconia frameworks are about 421–800 MPa and 6–8 MPa.m$^{1/2,}$ respectively. Among different

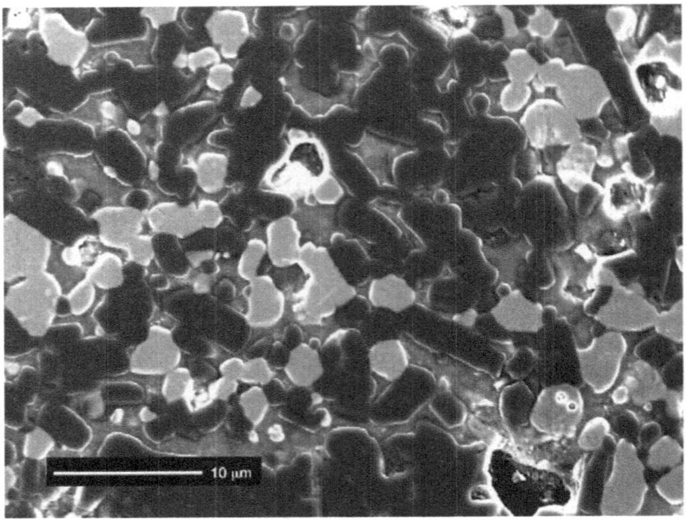

Fig. 3.17 Microstructure of glass-infiltrated Zirconia (In-Ceram Zirconia, VidentTM, Brea, CA) illustrating zirconia grains (bright particles) and alumina grains (dark particles). (Reprinted from Dental Materials, Denry & Kelly, State of the art of zirconia for dental applications, no. 302, Copyright (2008), with permission from Elsevier) [22]

glass-infiltrated ceramics, alumina/zirconia ones are highly opaque, similar to that of metal alloy cores, limiting their use for anterior regions. Translucent In-Ceram Spinell (VITA Zahnfabrik) composed of magnesia and alumina in the form of spinel ($MgAl_2O_4$) was introduced in 1994 to replace the opaque core of In-Ceram Alumina. This ceramic provides machinability with the CEREC inLab system (Sirona Dental Systems); however, the low flexural strength (350 MPa) of this system limited its use for posterior crowns [15].

3.1.5.3 Polycrystalline Solids of Alumina and Zirconia

Polycrystalline solids are solid-sintered, monophase ceramics formed by direct sintering of crystals using high sintering temperatures. These glass-free polycrystalline materials have great strength and toughness. However, the necessity of high temperatures has prohibited the use of these high-strength frameworks for crowns and FPDs. Solid-sintered, monophase ceramics could be manufactured using three methods. DCS Precident (DENTSPLYAustenal, York, PA) is the first method, which machines the framework from a solid-sintered block. This method is expensive, time-consuming, and requires manual adjusting of the coping. The second method is the Procera system, which utilizes an oversized die for applying a slurry of alumina or zirconia. The recently introduced method machines an oversized coping from a partially sintered block of zirconia material followed by sintering process during which it shrinks to match the die [9, 15].

Pure Alumina Core. Naturally, alumina (Al_2O_3) is present as corundum or emery, which can be transformed irreversibly to alpha-alumina (α-Al_2O_3) at temperatures above 1050–1200 °C [7]. Alumina by the side of zirconia is among the bioceramics widely used in dentistry due to its biocompatibility, inertness, good esthetics, favorable mechanical strength, and superior corrosion and wear resistance, which make the material suitable to be applied for prosthetic restorations (limited to small sizes due to high opacity), orthodontic implants, brackets, and high-stress regions. Clinically, the biocompatibility of alumina has been evaluated and recorded for more than forty years. However, it has a high melting point making it extremely hard to cast. On the other hand, the high hardness (20–30 GPa) of alumina ceramic is a major limitation against machining these ceramics. The stiffness of alumina is about 10 times greater than that of dentine, which limits their use where high elastic match between tooth structure and prosthesis is required. Pure alumina cores present greater strength and translucency compared to those of glass-infiltrated core ceramics. Procera AllCeram (Nobel Biocare, Stockholm, Sweden) is a densely sintered alumina-based ceramic introduced in 1993 by Anderson and Ogen composed of a pure and dense alumina core (99.9% Al_2O_3) veneered with a low-fusing feldspathic porcelain veneer. Exhibiting the highest flexural strength of alumina-based materials and fracture toughness equal to 487–700 MPa and 4.48–6 MPa.m$^{1.2}$, respectively, this product has been used as anterior and posterior crowns, veneers, onlays, inlays, short-span bridges, and implant abutments. The Procera AllCeram provides more translucency and strength compared to those of In-Ceram Zirconia. The production process of Procera AllCeram consists of the production of an enlarged die from an impression and digitization of the die via a computer and mechanical profiling device (CAD). The enlarged die is then utilized to dry-pressing the pure alumina on it. Afterward, following by the sintering of the oversized alumina, the desired shape and size of alumina core are achieved. The high-purity alumina powder is then compacted on the enlarged model and sintered at high temperature of 1550 °C to eliminate the pores and decrease the size. The fully sintered polycrystalline alumina cores present a negligible amount of the glass phase within their microstructure. At the final stage, the translucent feldspathic porcelain veneer with the suitable coefficient of thermal expansion is applied to the highly crystalline alumina core ceramic using sintering process. TechCeram (TechCeram Ltd, UK.) is another alumina-based core product sintered at 1170 °C and veneered with a conventional feldspathic ceramic used for anterior and posterior crowns and inlays. TechCeram is produced via spraying fine and pure alumina particles onto a refractory die, which has been produced in advance using impression and digitization. Spraying the alumina particle into an oxygen/acetylene flame leads them to melt onto the refractory die forming a core layer with a thickness between 0.3 and 0.4 mm. The core ceramic is then veneered using feldspathic porcelain to enhance its esthetics. This product exhibits a density of 80–90%, and flexural strength of about 300 MPa. TechCeram is suitable to be used as crowns and inlays of anterior and posterior regions [7, 13–15].

Pure Zirconia Core. Zirconia is an exceptional material due to its high flexural strength (900–1,100 MPa) and fracture toughness, which are almost twice the reported amounts for alumina-based ceramics. The introduction of zirconia ceramics has opened a wide range of all-ceramic applications and has slowly replaced alumina-based ceramics. Zirconia is an example of the modern dental ceramics generally used as core or substructure materials due to its high mechanical properties. Zirconia-based ceramics are recommended for FPDPs due to their high strength compared to that of alumina-based and lithium disilicate-based ceramics. Monolithic zirconia, zirconia-containing lithium-silicate ceramics, and interpenetrating phase composites are novel ceramics developed for restorative dentistry [9].

Tetragonal Yttria-Stabilized Zirconia Polycrystal. Tetragonal yttria-stabilized zirconia polycrystal (Y-TZP) is an important monolithic (full-contour) restoration, which has recently been introduced to overcome issues such as chipping of porcelain layers applied over zirconia. Zirconia ceramics exhibit various advantages including perfect mechanical properties, clinical performance, and facile machining with the help of computer-aided design and computer-aided manufacturing (CAD/CAM) technology. Zirconia is a polymorphic ceramic with three crystallographic forms at different temperatures, which are either monoclinic (below 1170 °C), tetragonal (between 1170 °C and 2370 °C), or cubic (between 2370 °C and 2680 °C). The reversible tetragonal-to-monoclinic phase transformation occurring while cooling below 1170 °C is accompanied by a 3–5% volume expansion followed by high internal stresses. The addition of yttrium oxide (Y_2O_3 3% mol) to pure zirconia could prevent this transformation and partially stabilize the tetragonal phase at room temperature. Yttrium oxide partially stabilized zirconia (Y-TZP) presents chemical and dimensional stability, superior mechanical properties, and fracture toughness higher than any other dental ceramics available due to the stress-inducement of tetragonal-to-monoclinic transformation that hinders crack propagation. In other words, the tensile stress created at a crack tip causes tetragonal-to-monoclinic transformation associated with 3–5% volume increase resulting in the formation of compressive stresses at the crack tip, which can queer the tensile stress and retard crack propagation. Tetragonal yttria-stabilized zirconia exhibits the highest values of flexural strength (900–1500 MPa) and fractures toughness (4–5 MPa.m$^{1.2}$) compared to other all-ceramic materials, and that is why this ceramic material is called *ceramic steel.* Translucent cubic zirconia with flexural strength values of 600–700 MPa followed by lithium disilicate-reinforced ceramics (262–420 MPa) occupy the second and third grades of ceramics with high strength. The modulus of elasticity and coefficient of thermal expansion for 3Y-TZP are about 210 GPa and 10.5×10^{-6} °C^{-1}, respectively. Y-TZP ceramics present high biocompatibility, low bacteria accumulation, metal-like radiopacity, and need smaller connector areas compared to other all-ceramic cores [3, 15, 19, 21] To achieve the desired mechanical and chemical properties, the utilized fabrication process should be precisely monitored. Blank fabrication, green machining, sintering process, and surface treatments are to be considered. The initial phase of the fabrication process, which is called blank fabrication, is highly dependent on the composition of powder, powder granulometry,

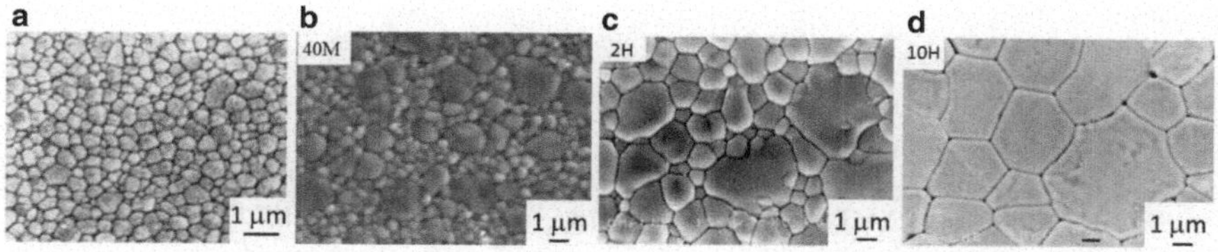

Fig. 3.18 SEM images of 3Y-TZP showing the effect of the duration of heat treatment on grain size **a** after sintering at 1450 °C; after heat treatment at 1650 °C for **b** 40 min, **c** 2 h, **d** 10 h. (Reprinted from Dental Materials, Turon-Vinas & Anglada, Strength and fracture toughness of zirconia dental ceramics, no. 367, Copyright (2018), with permission from Elsevier) [23]

particle size distribution, thermal treatment before sintering, and pressing process. The spray-dried 3Y-TZP agglomerates generally have diameters between 20 and 80 μm, which cause a coarse pseudo-grain structure, mostly occurring while using uniaxial press. Micropores present at the pseudo-grain boundaries may decrease resistance to low-temperature degradation (LTD). Sintering conditions including sintering temperature and duration time strongly influence the metastability, mechanical properties, and resistance to LTD of 3Y-TZP such as its microstructure, grain size, cubic phase amount, and yttrium segregation. Sintering temperature, heat treatment, and its duration directly influence the grain size in 3Y-TZP ceramics. As shown in Fig. 3.18, the grain size of 3Y-TZP increases from about 0.3 μm (sintering at 1450 °C) to more than 2.0 μm (heat treatment at 1650 °C for 10 h) with different durations of heat treatment. Occurring phase transformation is more probable for larger grains providing better mechanical properties; however, they decrease the resistance to LTD. On the other hand, the sintering temperature will also determine the amount of cubic phase and yttrium distribution, which influences resistance to LTD [19].

One of the challenges of the use of full-contour zirconia restorative materials is its color and translucency. Infiltration of metal salts at low concentrations is a way of coloring zirconia exhibiting some disadvantages including non-uniform color and limited diffusion depth of color. Different color shades of monolithic zirconia could be achieved by the addition of pigments using dip coating of the pre-sintered product or producing pre-colored blocks of monolithic zirconia manufactured from synthesized powders and pigments together. The latter approach provides much more homogeneous color shades. In order to provide better esthetics to opaque Y-TZP restorations, the addition of the oxides of iron, titanium, cerium, etc. brings about color shades similar to natural teeth (Fig. 3.19). To enhance the translucency of these ceramics microstructural modifications such as limiting alumina content, increasing density, decreasing grain size, addition of cubic zirconia, and decreasing the number of impurities and structural defects have been proposed and carried out. Grain size affects the translucency of zirconia restoratives in a way that smaller grains present lower translucency. In contrast, larger grains bring about smaller number of grain boundaries and decrease light scattering, which efficiently enhances the translucency of Y-TZP. As larger grains lower mechanical properties and stability of the tetragonal phase, the translucency cannot be enhanced by increasing the grain size. Another

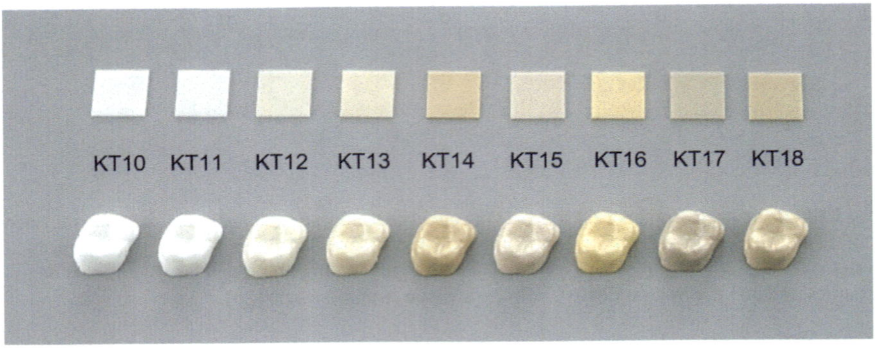

Fig. 3.19 Different color shades of Y-TZP restorations (Noritake). (Reprinted from Butterworth-Heinemann, Piconi et al., Alumina-and zirconia-based ceramics for load-bearing applications. In Advanced ceramics for dentistry, no. 240, Copyright (2014), with permission from Elsevier) [24]

solution proposed to increase the translucency of Y-TZP is to reduce the grain size in large extents reaching a critical value to reduce the birefringence phenomenon, which happens in Y-TZP because of the presence of more than 90% of tetragonal crystal phase. The anisotropic behavior associated with the difference of refractive index considerably scatters the light. Utilization of cubic zirconia providing optical isotropic behavior also helps to overcome these scattering effects and increases the translucency [19].

Lava™ (3MESPE, St. Paul, MN), DCS Precident (DCS Dental AG, Allschwil, Switzerland), and Cercon® (DENTSPLY, York, PA) are the most important products with Y-TZP frameworks used as single crown and three-unit bridges at the interior and posterior regions. In the Lava system, an optical system scans the die followed by designing and milling an enlarged partially sintered Y-TZP blank using CAD/CAM milling. The scanning process of this system lasts about 5 min for a crown and 12 min for a 3-unit FPDP. The milling process also takes about 35 min for a crown and 75 min for a 3-unit FPDP. Afterward, the milled product is colored and sintered. The framework of this system is translucent and can be produced in 7 different color shades (Vita-Lumin shade guide). Cercon (Dentsply Ceramco, York, Pa) is another CAD/CAM system based on the lost-wax method used for producing these restorations in which a white-colored framework is used. Similar to the previous Lava system, a partially sintered Y-TZP blank is used in this system with CAD/CAM technology to produce the framework. In DCS Precident (DCS Dental AG, Allschwil, Switzerland) system, a fully sintered, partially stabilized, fully sintered zirconia blank (95% ZrO_2 and 5% Y_2O_3) is used for CAD/CAM procedure [15]. Metastability and tribological behavior of monolithic zirconia restoration are associated with its final surface condition. Studies performed on various surface finishing techniques of monolithic zirconia have shown that polished surfaces provide less enamel wear of the antagonist surface compared to the glazed ones, and well-polished monolithic Y-TZP brings lower wear of the antagonist surfaces. Grinding and sandblasting of monolithic zirconia cause phase transformation followed by compressive stress on the surface. As a matter of fact, mirror finish polishing is essential besides heat treatment to eliminate compressive stresses formed in this stage by reversing the phase transformation. Glazed zirconia is more abrasive than polished or as sintered zirconia causing greater enamel cracks in natural teeth. Besides this, a massive formation of monoclinic phase decreases the wear resistance of 3Y-TZP due to the microcracking, larger grain size, and weakness of the *m*-phase. In parallel, increase of the 3Y-TZP grain size increases the wear resistance [19]. Aimed at improving the marginal adaptation of Y-TZP, improved CAD/CAM systems have been developed. The aging phenomenon of different types of monolithic zirconia is an important factor in the performance of these materials. The results of laboratory studies have illustrated that some brands do not age in the oral environment while others do show the tetragonal-to-monoclinic transformation. There are few evaluations performed on the utilization of monolithic zirconia crowns. In one of these studies performed on 60 patients, 82 monolithic zirconia crowns were installed. In this manner, 6 patients faced some difficulties after 3 years due to retention loss of crowns and endodontic complications. Another study comprised of five years data collected from 39,827 restorations including 1,952 anterior single crowns, 29,808 posterior single crowns, 1,779 anterior fixed dental prostheses, and 6,288 posterior fixed dental prostheses. The resulted data have indicated that 0.97% of anterior single crowns, 0.71% of posterior single crowns, 3.26% of the anterior fixed dental prostheses, and 2.42% of the posterior fixed dental prostheses were subjected to catastrophic fractures [17].

Zirconia-Toughened Alumina (ZTA) Cores. ZTA ceramics are composed of 70–90 wt% alumina and 10–20 wt% zirconia. The toughening mechanism of the ZTA ceramics is similar to that of Y-TZP ceramics. In the microstructure of ZTA, stress-induced tetragonal-to-monoclinic transformation of zirconia happens to cause the crack formation in the structure of zirconia followed by their phase transformation resulting in the increase of zirconia particle numbers, which induces compressive stresses within the structure of alumina. During this process, the strength and fracture toughness of alumina increases. The flexural strength and the fracture toughness of ZTA are about 760 MPa and 5–7 $MPa.m^{1/2}$, respectively. On the other hand, these values are about 330 MPa and 3.5 $MPa.m^{1/2}$, respectively for alumina ceramics [1].

3.1.6 Dental Ceramics Fabrication

Dental restorations can be produced via various technologies depending on the used material, desired shape, and the application of restorative. Additive manufacturing, sintering, cast and sintering, heat pressing, slip casting, and computer-aided design/computer-aided manufacturing (CAD/CAM) techniques are among the processing techniques of dental restoratives [4].

3.1.6.1 CAD/CAM Techniques

Primary machining methods were based on the application of a resin composite to make a replica of the restoration, which is then scanned via a scanning tool of the scanning system. The restoration is then fabricated from a ceramic block using a copy-milling machine. Nowadays, the CAD/CAM (computer-aided design/computer-aided manufacturing) technique has taken the place of copy-milling. The CAD/CAM technology was first introduced in dentistry by Duret in the early 70 s. This technology was initially introduced for machining fully sintered ceramic blocks. The developments of fabricating reliable all-ceramic restorations with superior properties are in debt of the novel CAD/CAM systems, which have been used for more than 30 years in dentistry. Nowadays, this technique can be used with various materials such as veneering ceramics, Y-TZP ceramics, resin composites, and metal alloys and is applicable to mill inlays, onlays, veneers, and crowns in a very short time and one clinical appointment [1, 8, 12].

Through this technique, an optical laser scanner is used to scan the replica of dental restoration and digitalize the form and dimensions of the coping (framework) and transfer the data to the computerized numeric control (CNC). Later, using special software of the chairside CAD/CAM system (3 M True Definition Scanner, 3 M Company; CEREC AC, Sirona Dental Systems, LLC; PlanScan Restorative System, Planmeca; iTero Intra Oral Digital Scanner, Align Technologies), the restoration is designed followed by its fabrication by diamond milling a sintered or pre-sintered ceramic blank (Fig. 3.20) [3].

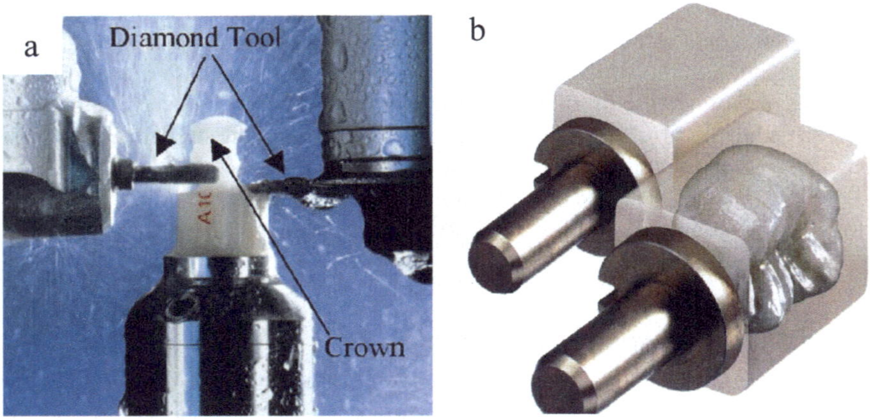

Fig. 3.20 Manufacturing procedure of a restoration via CAD/CAM technique in a Cerec system: milling process of the CAM (**a**), the machined crown from a blank (**b**). (Reprinted from International Journal of Machine Tools and Manufacture, Yin et al., An overview of in vitro abrasive finishing & CAD/CAM of bioceramics in restorative dentistry, no. 1020, Copyright (2006), with permission from Elsevier) [25]

The CAD/CAM technique involves two different methods in dentistry applications, which are either machining of the prosthetic restoration from a sintered block (hard machining) or machining a partially sintered or green block (soft machining) followed by a final heat-treating step to achieve fully sintered products. The soft-machining technique requires enlarged blocks to overcome followed shrinkage and is more suitable for fabricating hard ceramics such as alumina and zirconia. Milling the soft pre-sintered or green blocks reduces the milling time as well as decreasing the wear of milling devices compared to the fully sintered blocks. Normally, the coping presents a linear shrinkage of 20–25% while sintering. In the Lava (3 M ESPE, St. Paul, Minn) method, a contact-free optical device scans the die (5 min for a crown and 12 min for a 3-unit FPDP) and the CAD software designs an oversized framework, which is milled from green or pre-sintered blanks. The milling process takes 35 min for a crown and 75 min for a 3-unit FPDP. After selection between the seven different color shades possible for the framework, it will be sintered in a special oven for 8 h. In the method of Cercon (Dentsply Ceramco, York, Pa), which is used for designing and milling of zirconia, the conventional waxing technique is used to design the framework. DCS Precident (DCS Dental AG, Allschwil, Switzerland) on the other hand, uses fully sintered DC-Zirkon ceramic, which consists of 95% ZrO_2 that is partially stabilized with 5% of Y_2O_3. Denzir (Decim AB, Skelleftea, Sweden) is another method, which designs and mills ceramic inlays from Y_2O_3 partially sintered zirconia blocks. The advantages of the method using fully sintered blocks are high precision of the shape of products, less time-consuming, and no need for heat treatment. Nevertheless, machining high-strength materials leads to the wear of the machining tools. Besides this, brittle materials are susceptible to the formation of microcracks and surface defects while machining. On the contrary, microcracks formed during machining of the materials fabricated using partially sintered blocks are probable to be eliminated after the later sintering process. However, the final sintering may lead to dimensional changes. The main weakness point of the CAD/CAM systems is the great material waste associated with the fabrication process. Aimed at improving this disadvantage of the CAD/CAM technique, additive manufacturing systems (solid free-form fabrication) have been developed [1, 3, 15].

Hard-Machined Ceramic Restorations. Machinable fluormica glass-ceramics were the first ceramics presenting great machinability characteristic due to the presence of interlocked "house-of cards" microstructure and cleavage planes of mica crystals. These days, feldspar-based, leucite-based, and lithium disilicate-based ceramics are examples of machinable dental ceramics suitable for hard machining. Introduction of partially crystallized lithium-silicate ceramics caused a great advance in the development of hard-machining technique. The partially crystallized ceramics include lithium metasilicate (Li_2SiO_3) crystals and lithium disilicate ($Li_2Si_2O_5$) crystal nuclei, and are machined easier than fully crystallized ones. The pre-treatment process of these ceramic blocks dictates their translucency. Ceramic blocks containing fewer and larger pre-crystallized lithium metasilicate are called high translucent (HT) ceramics while blocks containing smaller crystals are called low translucent (LT) ceramics. Full crystallization of HT and LT ceramics occurs after a heat treatment at 850 °C for 10 min. The microstructure of heat-treated HT ceramic is composed of layered lithium disilicate crystals dispersed in a glassy matrix accompanied with spherical lithium phosphate crystals. On the other hand, LT ceramic comprises a great density of small lithium disilicate crystals interlocked together as well as lithium phosphate crystals after heat-treating process. The latest machinable ceramic produced commercially is a lithium-silicate glass-ceramic with 10 wt% addition of zirconia (ZLS) with great mechanical properties and translucency. The final microstructure of machinable ZLS ceramics is composed of fine lithium metasilicate and larger lithium disilicate crystals interlocked together [1, 3, 12].

Soft-Machined Ceramic Restorations. It was in 2001 when the first partially sintered zirconia ceramics were used for soft machining of dental restorations. These ceramics are easy-milled leading to a decrease in milling time and tool wear. The used material is a partially stabilized (by 3 mol% of yttrium) polycrystal of tetragonal zirconia (3Y-TZP). Due to the great shrinkage of this ceramic during sintering (20%–25%), the designed and machined restorations should be enlarged. The microstructure of the final dental restorations is composed of densely packed tetragonal zirconia grains with grain sizes ranging between 0.2 and 1.0 μm. These products provide the highest flexural strength (900–1500 MPa) and fracture toughness (about 6 $MPa{\cdot}m^{0.5}$) compared to all other dental ceramics produced commercially and are used for single and multi-unit restorations of anterior and posterior regions. Slight mechanical properties and microstructural variation of 3Y-TZP products are due to different sintering temperatures and sintering durations applied by different manufacturers. Generally, sintering temperatures applied by different manufacturers range between 1350 °C and 1550 °C and the duration time of sintering is between 2 and 6 h. Manipulation and application of 3Y-TZP ceramics by dentists or technicians is tricky and sensitive to some extent due to the possibility of occurring tetragonal-to-monoclinic transformation while grinding or air abrasion. 3Y-TZP dental restorations accompanied with soft-machining technique, present great clinical success and the best mechanical properties of all other ceramic materials currently available [1, 3, 12].

3.1.6.2 Additive Manufacturing Techniques

Although in the next coming chapters of this book, additive manufacturing (AM) techniques will be discussed in detail but because of the critical role of ceramics as a dental material, in here, a view has been taken. Additive manufacturing systems or 3D-printing technologies (solid free-form fabrication) have been developed to overcome the disadvantages of previous processing methods of dental restorations. Selective laser sintering (SLS) or melting (SLM), direct 3D-printing, and stereolithography (SLA) are among the different fabrication methods based on additive manufacturing. In SLS/SLM techniques, a high-power laser beam sinters selected parts of a powder layer. This technique is widely used for the fabrication of metal alloys. Direct 3D-printing is a manufacturing technique in which a ceramic suspension is directly applied resulting in the fabrication of complex shapes with high resolution. SLA is a very promising technique commonly used for the fabrication of complex ceramic shapes. In this technique, a suspension of ceramic particles and resin components (acrylates or epoxy monomers) are used. The polymerization of the resin part during printing is in charge of providing the shape of the product and will be vaporized during the sintering process of the ceramic. Opposite to the CAD/CAM techniques, the most important advantage of AM techniques is their low material waste. Hence, the disadvantage of these methods is the rough surface and poor marginal precision of the products [17]. Slurry additive manufacturing, also called robocasting, is a fabrication method of ceramics in which extruded filaments additively construct desired models. This method has been used for the fabrication of prostheses cores made of zirconia/alumina and orthopedic scaffolds [19].

3.1.6.3 Slip-Casting and Glass-Infiltration Fabrication

Slip-cast ceramics for dental restorations were introduced in the 1990s. In the slip-casting method, the substructures are fabricated using layer-by-layer application of a mixture of fine-grained Al_2O_3 powder and de-ionized water (condensation of aqueous ceramic slurry) on refractory gypsum die possessing several porosities. The water content of the porcelain slurry is absorbed using gypsum die. The firing process of the layered gypsum die results in the shrinkage of the die more than the condensed surface layer, which facilitates the process of separating them. The alumina layer consisting of almost 85% Al_2O_3 is then heated to 120 °C to remove the residual water content. The heating process continues slowly until reaching 1120 °C at which it stays for 2 h. The product of this process is a porous (5 vol.%) and interconnected structure, which is infiltrated with molten lanthanum aluminosilicate ($LaAl_2O_3SiO_2$) glass. The substructure product is then heated to 1100 °C for 4–6 h leading to diffusion of the melted glass into the pores. Products manufactured using this method present low porosity and defects as well as high fracture toughness. Zirconia and spinel can be also fabricated using this method with flexure strength equal to 300 MPa for spinel and 700 MPa for zirconia ceramics. The microstructure of slip-cast alumina contains large and irregular alumina grains resulting in the opacity of this product. Slip-cast spinel ceramics, on the other hand, present great translucency. Slip-cast zirconia-toughened alumina (34 vol.% alumina and 33 vol.% of 12 mol% ceria-stabilized zirconia) contains about 8 vol.% of porosity. The microstructure comprises of large alumina grains accompanied with smaller 12Ce-TZP grains and visible pores. These alumina-zirconia slip-cast ceramics provide the highest flexural strength and fracture toughness compared to all other slip-cast ceramics due to the presence of two toughening mechanisms, which are because of the stress-induced transformation of zirconia grains and crack deflection owed to the large alumina grains [1, 8, 13].

3.1.6.4 Layering and Sintering

Layering and sintering is a fabrication method consisting of applying a slurry made of ceramic powder and liquid onto a framework followed by its pre-heating in order to evaporate the liquid content and achieve a layer with the lowest porosity. In the next step, the layered framework is placed in a furnace under vacuum conditions. The sintering process occurs at high temperatures of around 900 °C. Firing ceramic powders at high temperatures is commonly used for metal-ceramic restorations. Employment of appropriate size distribution of powders providing high packing factor helps to the elimination of formed pores and leads to the formation of dense fabricated materials. Densification and grain growth are the two-effective heat-activated stages of the sintering process. As the final product shrinks about 40% during its sintering process, the green product should be fabricated larger than its desired dimension [8].

Microwave sintering is a method, which provides volume and surface heating in addition to enhancing atomic diffusion by lowering the activation energy, which leads to a decrease of the sintering time and sintering temperature. As a conclusion, microwave sintering generally leads to the formation of high-density materials with uniform grain size. There are some issues to be addressed in this process including expensive applicators with high frequencies, possibility of facing material overheating, and the formation of non-uniform microstructure due to temperature gradients created because of the volumetric heating. Nowadays, by using hybrid techniques such as a combination of microwave and infrared heating or the use of microwave-absorbing materials (e.g. SiC) these problems have been limited. Microwave sintering increases material productivity and reduces energy costs. Moreover, for example, microwave-sintered zirconia provides smaller grain size and increased translucency compared with those from conventional firing [19].

Mainly due to the success of machined and heat-pressed restorations, sintered all-ceramic systems are now mostly obsolete. Two main types of all-ceramic materials, which are available for the sintering technique include alumina-based ceramics and leucite-reinforced ceramics. Leucite-reinforced ceramics containing up to 45% by volume of tetragonal leucite were used for the fabrication of all-ceramic sintered restorations. Both sintered and slip-cast all-ceramic restorations have now been replaced by heat pressed or machined all-ceramic restorations containing similar crystalline phases with better-controlled processing steps [3].

3.1.6.5 Spark Plasma Sintering (SPS) (kinetic Engineering)

Controlling the kinetic in the sintering process is a crucial factor in the fabrication of nano-ceramics and composite-ceramics. Spark plasma sintering increases surface diffusion along with decreasing grain boundary diffusion. The setup used for fabricating materials using SPS consists of a powder container die, which is electrically conductive. A uniaxial pressure accompanied by strong direct current (DC) is then applied to the die providing rapid sintering at low temperatures. This method fails to fulfill the requirements of producing complex shapes of dental restoration [19].

3.1.6.6 Heat Pressing

The heat-pressing method for producing dental restorations was first introduced by Wohlwens and Scharer in 1990 and was used later by Ivolcar Vivadent (Liechtenstein). Heat pressing relies on the application of the lost-wax method and external pressure at high temperatures to sinter and shape the ceramic restoration. Heat pressing is used in dentistry to produce all-ceramic crowns, inlays, onlays, veneers, and more recently, FPDs. Authentic (Ceramy, D), Super Porcelain EX-3 Press (Noritake, JP), Ceramco Press (Dentsply, USA), In-Line PoM (Ivoclar Vivadent, FL), and IPS InLine are the examples of heat-pressed products. In the first stage of producing dental restorations via the heat-pressing method, a wax model of the restoration is precisely designed and constructed followed by placing the wax model in a phosphate-bonded refractory die. By burning the wax model, the formed empty mold is used to press the glass-ceramic or ceramic ingot into it. The heat-pressing temperature is chosen near the softening point of the ceramic. A pressure of 0.3–0.4 MPa is applied to the plasticized ingot block at high temperatures under vacuum to press it into the created mold. Heat pressing provides products with enhanced strength and low porosity content due to the high crystallinity, and small crystal size of the products. On the other hand, this procedure is simple, accurate, and presents superior marginal fits of the restorations. Dental technicians are usually familiar with this technique, commonly used to cast dental alloys. In addition, the equipment needed to heat-press dental ceramics is relatively inexpensive. The first generation of heat-pressed dental ceramics contains leucite as reinforcing crystalline phase (between 35 and 45 vol.%) and the heat-pressing temperature ranges between 1150° and 1180 °C. These products contain 9 vol.% porosity with a flexural strength of 120 MPa. The leucite-reinforced ceramics present great esthetics but due the lack of mechanical strength, their application in the single-unit restorations of anterior regions is limited. The second generation of heat-pressed dental ceramics contains lithium disilicate crystalline phase (about 65 vol.%). In the heat-pressing process of these products, the temperature changes between 910 °C and 920 °C. These products present porosity contents of about 1 vol.% and flexural strength of 350 MPa. The crystal alignment through heat-pressing process of lithium disilicate glass-ceramics can cause different mechanical properties in directions parallel or perpendicular to crystal alignments [1–3, 8, 13].

3.2 Cements

3.2.1 Introduction

Dental cements, also called luting agents, are adhesives and generally used in the final step of clinical procedure in order to fill the space between indirect restorations and prepared tooth along with making a bond between them during the desired time. The cementation or polymerization process occurring in these materials, which is known as the setting process, creates a fit bonding between the tooth and the restoration, which retains the latter in its place for an indefinite time and makes an impenetrable seal between the tooth and the adjoint restoration [26]. Throughout the years, several types of dental cements have been used in dentistry applications. Sorel in 1855 introduced the first acid-based cements, zinc oxychloride cements, prepared by mixing zinc oxide powder with a solution of zinc chloride. Afterward, in 1858 Feichtinger made efforts to use these cements in dentistry, which did not show any success [27]. The primary dental cements (conventional cements) were actually luting agents with the major function of filling the spaces between the prepared tooth and indirect restoration to fix them together using mechanical interlocking and friction. These primary cements did not have any adhesive properties. Zinc phosphate and zinc oxide-eugenol cements are among the first luting cements used in dentistry [28]. In 1955 for the first time in dental cements history, Buonocore used phosphoric acid as an etchant to create a porous enamel structure resulting in strong micromechanical bonding between the porous enamel and the used resin. Besides, the results of a research performed by Smith in 1968 lead to the introduction of polycarboxylate cements with the ability to form chemical bond to the tooth tissue based on ionic attraction between carboxyl groups (COO^-) present in the cement's structure and calcium ions (Ca^{++}) present in enamel and dentine [27]. In the past few years, dental cements with enhanced mechanical and physical properties have been used to better retain the newly developed dental restorations and fulfill the desired requirements for dental cements. Resin hybrid cements and resin cements with adhesive properties are the latest types of bonding agents introduced to dentistry applications holding modified and required properties [28]. Figure 3.21 shows a schematic of the development of dental cements from 1855 until today.

In dentistry, the principal function of dental cements is to attach the tooth tissue to the restoration material. This function could be achieved through three mechanisms. The first mechanism is the micromechanical interlocking occurring between the rough surfaces of acid-etched teeth, the cements, and the restoration. In fact, cements-luting mechanism is much more based on non-adhesive, micromechanical retention accompanied by molecular adhesion (resulted from bipolar van der Waals forces and weak chemical bonds between cements and tooth). This mechanism is the principal basis of retention for dental cements. Air abrasion and acid etching enhance micromechanical bonding by increasing surface irregularities of these cements especially while applying resin and resin-modified glass ionomer cements [26]. The second mechanism of bonding is based on the chemical adhesion

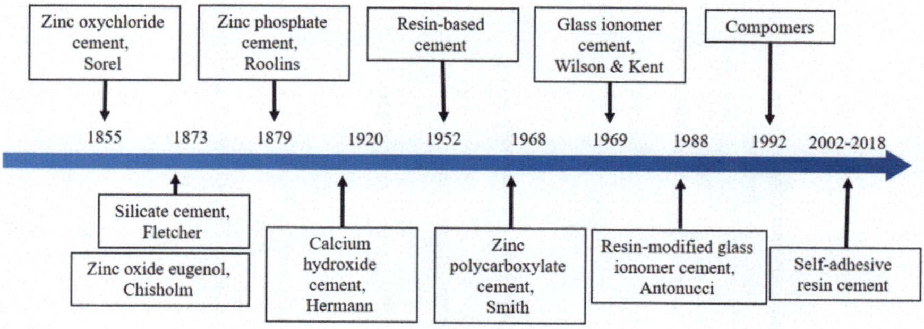

Fig. 3.21 A schematic of the development of dental cements from 1855 until now [27, 29–31]

to enamel or dentine resulted from hydrogen bonds or ionic and covalent chemical bonds between cements and teeth using coupling agents. Glass ionomers and resin-modified glass ionomer cements are able to form chemical bonds with either dentin or enamel. The third mechanism is based on wetting, penetrating, and formation of a bound layer, which occurs in novel dental bonding agents [32]. Bonding agents such as zinc polycarboxylate, glass ionomer, resin-modified glass ionomer, and self-adhesive resin cements are capable of making a chemical bond with the hydroxyapatite present in the tooth structure [3]. Dentin and enamel are consisting of about 50% and 97% hydroxyapatite, respectively [32]. Besides the major function of dental cements, which is to maintain the various types of restorations (e.g. alloys, ceramic crowns and bridges, inlays, onlays, and veneers) in their place during a desired period, they can be used in several other applications. Permanent cementation could be achieved using glass ionomer, resin-modified glass ionomer, and resin cements. On the other hand, thick consistencies of some dental cements are used as temporary fillings, thermal insulators, and mechanical supports. Dental cements have also great applications in endodontic, orthodontic, periodontics, and surgical dentistry as filling materials, root canal sealers, cavity interior bases, and cavity liners [33]. Among the various types of dental cements, zinc phosphate, glass ionomer, and zinc oxide-eugenol (ZOE) cements are suitable base cements for insulating deep cavities as a barrier against chemical and thermal injuries that are likely to happen to the pulp and tooth tissue. Then the restorative filling material could be seated over the cements base. The eugenol present in the composition of ZOE cements has sedative and painkilling effects, which makes them a good choice in several dentistry applications. The fluoride releasement capability and chemical adhesion of glass ionomer and compomer cements to the tooth have also made these materials useful in cementing and base applications. In addition, owing to the high strength of resin cements and their ability to bond to enamel and dentin, these cements are widely used as orthodontic cements and for cementing all-ceramic veneers, crowns, inlays, and resin-bonded bridges [3]. Table 3.5 gives a precisely defined summary of different possible applications of dental cements in dentistry and the fitting examples of cements materials.

Table 3.5 Summary of different applications of cements in dentistry [33–35]

Type of cements	Application	Cements material
Permanent luting cements	Cementation of cast alloy crowns and bridges	Adhesive resin (dual-cured), bioceramic, glass ionomer, resin-modified glass ionomer, self-adhesive resin, zinc polycarboxylate, zinc phosphate
	Cementation of all-ceramic or indirect composite inlays and onlays	Adhesive resin (dual-cured), bioceramic
	Cementation of zirconia-based and all-ceramic crowns and bridges	Adhesive resin (dual-cured), bioceramic, glass ionomer, resin-modified glass ionomer, self-adhesive resin
	Cementation of ceramic veneers	Esthetic resin (dual-cured or light-cured)
	Cementation of resin-bonded bridges	Adhesive resin (dual-cured)
Temporary luting cements	Temporary cementation of cast crowns and bridges	Zinc oxide–noneugenol, temporary resin
	Temporary fillings	Resin-modified glass ionomer, zinc polycarboxylate, glass ionomer, zinc phosphate
Low-strength base	Providing a protective action to the pulp from irritants or serve therapeutically as pulp-capping agents	Calcium hydroxide (self-cured and light-cured), glass ionomer, resin-modified glass ionomer, zinc oxide-eugenol
High-strength base	Providing thermal protection for the pulp and mechanical support for restorations	Compomer, glass ionomer, resin-modified glass ionomer, zinc polycarboxylate, zinc phosphate, zinc oxide-eugenol
Cavity liner	Providing a barrier against fluid penetration and therapeutic effects	Calcium hydroxide in a suspension
Cavity varnish	Providing a barrier against fluid penetration into the underlying dentin, reducing postoperative sensitivity and reducing tooth discoloration	The resin in a solvent, resin-modified glass ionomer
Special application cements	Retention of orthodontic bands	Glass ionomer, resin-modified glass ionomer, resin, Zinc polycarboxylate, zinc phosphate
	Direct bonding of orthodontic brackets	Resin
	Root canal sealer	Zinc oxide–eugenol
	Direct pulp capping	Tri-calcium silicate cements (ProRoot MTA, Biodentine)

3.2.2 Requirements for Choosing the Best Cements

The physical requirements of dental cements are specified in the standards from the International Association for Standardization. In this manner, ISO 3107:2004 describes the physical requirements for ZOE and noneugenol cements, ISO 9917-1:2007 describes it for powder/liquid acid–base cements, ISO 9917-2:2010 represents those requirements for resin-modified cements, and ISO 4049:2009 presents it for polymer-based cements [3]. To fulfill the demands of cements in dentistry applications, the used cements should prove various properties in general and in particular according to the specified application. Generally, an acceptable dental cement must represent a biocompatible response in the body and do not show any harmful effects on the tooth and neighboring tissue as well as providing appropriate mechanical properties after setting especially for those, which are used as permanent cements. Good adhesion to the tooth and restorative material, low shrinkage during setting, and low viscosity are also important to easily manipulate the cements and prevent penetration of oral fluid and bacteria to the tooth tissue and underlying pulp. However, radiopacity is another important factor allowing to make a difference between the cements and developed caries. In spite of the fact that various types of cements have been introduced for different dentistry applications, finding an ideal dental cement with superior properties is still a challenge for dentists and researchers [3, 31]. Table 3.6 illustrates the requirements of an ideal cement in detail including material examples. After all, selecting a convenient material to be used in dentistry applications is a crucial factor for providing a long-term and successful function. However, by considering the capabilities and deficiencies of different proposed cements and evaluating them, it is possible to select an appropriate cements material for the specified application [3, 31]. Table 3.7 illustrates different functions and requirements of cements materials used for distinct dentistry applications.

Choosing the most appropriate and applicable cements for a certain application does stand not only on the strength of the restoration material (weaker materials need to be cemented with stronger cements) but also the need for retention for the restoration. Figure 3.22 illustrates the best choice of cements for different applications [28].

3.2.3 Classification of Dental Cements

Different classifications of dental cements have been reported in the pieces of literature. Craig [3] classified dental cements according to their clinical usage into definitive and provisional (temporary) cements based on the required function time of cements. Definitive cements are used for permanent bonding of restoration to tooth for the longest time possible while provisional cements fix the restoration for shorter times aiding to finish the permanent restoration. In this manner, zinc oxide-eugenol (ZOE), zinc oxide noneugenol and calcium hydroxide cements are

Table 3.6 Ideal properties of dental cements and material examples [3, 31, 36]

Property	Requirements of ideal cements	Effect	Perfect cements	Imperfect cements
Biocompatibility	High biocompatibility, Non-allergic and non-carcinogenic, Non-toxic and non-irritant, Harmless to the pulp	Protecting pulp tissue from external stimuli	Zinc oxide, Polycarboxylate	Low pH acid–base cements, Self-etch resin, RMGI, Resin-based
Interfacial sealing and anti-cariogenic activity	Perfect sealer of the tooth/restoration interface	Preventing the formation of secondary caries caused by bacteria penetration, Preventing excessive fluid movement in the dentinal tubules that may cause hypersensitivity	Fluoride-containing cements, Glass ionomer, RMGI	–
Caries inhibition	Fluoride releasement, Antimicrobial properties	Avoiding plaque accumulation and physical exclusion of microorganisms from the exposed dental tissues and the interface	Glass ionomer, RMGI, Compomer	–
Film thickness	Low film thickness (<25 μm)	Allowing proper seating of the restoration	Zinc phosphate, Glass ionomer, Polycarboxylate, RMGI, Self-adhesive resin-based	Resin-based, Compomer
Manipulation	Long working time, Short setting time after restoration seating, Low dimension or temperature changes during setting	Properly seat the restoration, Ease and accuracy of mix and use	Compomer, RMGI, Resin-based, Self-adhesive resin-based	Zinc phosphate, Polycarboxylate, Glass ionomer

(continued)

Table 3.6 (continued)

Property	Requirements of ideal cements	Effect	Perfect cements	Imperfect cements
Excess material removal	Ease of excess material removal	Easy cleanup	Zinc phosphate	Resin-based, Self-adhesive resin-based, Compomers, RMGI
Wear resistance	High wear resistance	Decreasing susceptibility to wear when margins are located farther from the gingival area and approach the occlusal area	Composite resin	–
Compressive strength	High	Supporting tooth structure and providing clinical performance	Self-adhesive resin-based, Resin-based, Compomer	Zinc phosphate, Polycarboxylate, ZOE, EBA-reinforced ZOE
Elastic modulus	Equal to dentine (13.7) or between tooth and restoration	Decreasing transfer loads to the tooth and stress distribution	Zinc phosphate, Glass ionomer	Polycarboxylate, RMGI
Fracture toughness	High fracture toughness	Resist crack propagation that may cause breakage	Composite resin	Glass ionomer, RMGI
Solubility	Resistant to dissolution	Affecting marginal integrity of the indirect restoration and decrease plaque accumulation	Resin-based, Compomer, Glass ionomer, RMGI, Self-adhesive resin-based	Zinc phosphate, Polycarboxylate, ZOE, EBA-reinforced ZOE
Esthetics	Opaque cements are suitable for dark/metal/discolored backgrounds	Affecting the background effect	Zinc polyacrylate (opaque) for zirconia ceramics on titanium backgrounds	ZOE and zinc phosphate (opaque) for zirconia ceramics on tooth-colored backgrounds
	Tooth-colored cements are a good choice for tooth-colored backgrounds	Affecting the background effect	RMGI for monolithic zirconia crowns	Resin-based (tooth-shaded) for zirconia ceramics on titanium backgrounds

(continued)

Table 3.6 (continued)

Property	Requirements of ideal cements	Effect	Perfect cements	Imperfect cements
Radiopacity	Higher than dentine and radiopaque	Allowing radiographic diagnostic to detect eventual gaps, secondary caries, overfilling, or underfilling	Glass ionomer, Composites of zinc oxide, calcium oxide, and barium glass, RMGI	Quartz, lithium aluminum glasses, and silica fillers are not radiopaque
Microleakage	Resistant to microleakage	Decreasing pulpal response, hypersensitivity and endodontic failure	Resin-based, RMGI, Self-adhesive resin-based	Zinc phosphate, Polycarboxylat, Compomer, ZOE, EBA-reinforced ZOE
Adhesion	Micromechanical interlocking or friction	Reducing the risk of displacements Reducing the risk of restoration fracture Improving interfacial sealing	Resin-based, Self-adhesive resin-based, RMGI, Compomer	Zinc phosphate
	Molecular adhesion (chemical or physical bonding)		Polycarboxylate, glass ionomer, RMGI, compomer	
Shelf life	Without a change of viscosity during time	Providing a convenient period during which physical and mechanical properties maintain unchanged and minimize waste of materials	Zinc phosphate	Glass ionomer
Cost	Low	Economic	Zinc phosphate, Zinc carboxylate, Glass ionomer	Resin composite, RMGI

Remarks RMGI = resin-modified glass ionomer, ZOE = Zinc oxide-eugenol, EBA = Ethoxybenzoic acid

Table 3.7 The specified functions and requirements of cements materials used for distinct dentistry applications [32, 34]

Material function	Requirements	Applications
Cavity liner	Thermal barrier	Protecting the pulp from sudden changes in temperature especially with metallic fillings
	Chemical barrier	Forming protective barrier against potential chemical irritants
	Electrical barrier	Preventing the formation of galvanic cell which accelerates corrosion and causes pain after operation
	Flowable	Facilitating the manipulation process and allowing sufficient time for mixing and placing the cements in cavity
	Radiopaque	Aiding the diagnosis of caries around the filling
Luting cements	Adequate setting time	Allowing sufficient time for mixing the initial materials, applying the cements and seating the restoration in its place
	Low initial viscosity	Allowing the flow of the cements (pseudo-plasticity) and proper seating
	The thermal and electrical barrier	Avoiding harmful effects of alloy restorations
	High retention	Giving mechanical resistance to the displacements of restoration and resisting fracture formation
	Low soluble	Preventing dissolution and erosion of the cements, which can cause failure by loss of retention or by the initiation of caries in the tooth substance
Endodontic cements (root canal sealer)	Flowable	Allowing the cements flow and be pressed into the complex structure of the root canal
	Biocompatible	Avoiding pulp inflammation and harmful effects on the tooth and surrounded tissue
	Sealer	Forming a good seal along the walls and at the apex of the canal preventing bacteria accumulation
	Radiopaque	Allowing the dentist to confirm the attainment of a satisfactory root filling radiographically
	Adequate setting time	Enabling material to be forced into the root canal before setting
	Adequate working time	Allowing the completion of the procedure and checking radiographs to ensure the satisfaction of root filling
Orthodontic cements	Easy to remove excess	Leaving a minimum amount of residual cements attached to the teeth after removing the brackets and bands
	Antibacterial	Avoiding plaque accumulations and decalcification of the enamel around brackets
	Low viscosity	Enabling placements of the bracket and creation of a thin film of cements

(continued)

Table 3.7 (continued)

Material function	Requirements	Applications
	Pseudo-plastic	Preventing drift in the position of the bracket when placements are completed
	Rapid setting	Helping to ensure that the position of the bracket does not change (e.g. through light activation)

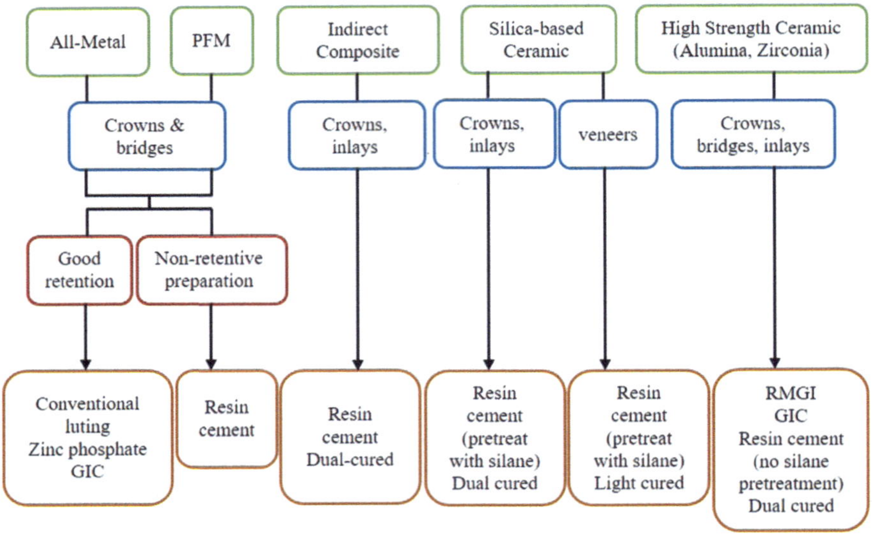

Fig. 3.22 Best selection of cements for different applications. Redrawn from Sunico-Segarra et al. [28]. *Remarks* PFM = Porcelain-Fused-to-Metal, GIC = Glass Ionomer Cements, RMFI = Resin-Modified Glass Ionomer

classified as temporary cements. On the other hand, definitive cements are divided into low (zinc phosphate, silicate cements), medium (polycarboxylate cements), and high (glass ionomer cements) adhesive potential cements [37]. Furthermore, Sunico-Segarra et al. [28] divided dental cements into two groups: the conventional luting cements (e.g. zinc phosphate, polycarboxylate, glass ionomer, and resin-modified glass ionomer) and the bonding cements (resin cements). Moreover, Powers and Wataha [33] classified cements as water-based (e.g. glass ionomer cements, resin-modified glass ionomer cements, zinc polycarboxylate, and zinc phosphate), oil-based (zinc oxide-eugenol), and resin-based (adhesive resin, self-adhesive resin, and esthetic resin) cements. O'Brien has also categorized dental cements based on the type of matrix bond, which are phosphate-bonded, phenolate, polycarboxylate, resin and resin-modified glass ionomer cements. The most well-known classification of dental cements was performed by Wilson based on the predominant setting mechanism of components, which are acid–base reaction occurring in acid–base cements

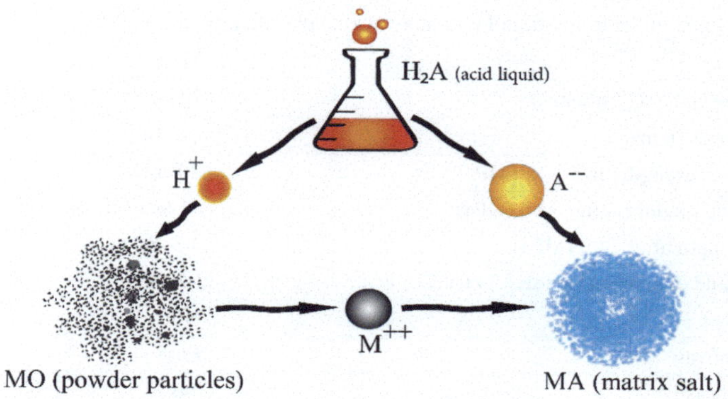

Fig. 3.23 Schematic of an acid–base setting reaction. Redrawn from Wilson et al. [39]

and polymerization reaction occurring in resin-based cements. Glass ionomers (GI), resin-modified glass ionomers (RMGI), zinc oxide-eugenol (ZOE), zinc polycarboxylate, ethoxybenzoic acid (EBA)-reinforced ZOE cements and zinc phosphate are some examples of acid–base cements. On the other side, composite resin cements, compomers, and self-adhesive resin cements are examples of some important resin-based cements [3, 38]. Acid–base cements (AB) are formed by mixing powder (metal-oxide or silicate as a cation releaser or proton acceptor) and an acid (proton donor) together resulting in the formation of a complex cements matrix (salt) including the unreacted particles of powder. The chemical reaction of the cementation process could be represented as below:

$$\text{MO (base)} + \text{H}_2\text{A (acid)} = \text{MA (Salt hydrogel)} + \text{H}_2\text{O}$$

The setting mechanism appearing in AB cements is a consequence of gelation followed by cation releasement from the oxide or silicate and the interaction (ion bonding and change of hydration state) of the cations with acid. The application of different acids and bases for acid–base cements generates cements systems with distinct properties. Figure 3.23 shows a schematic of the acid–base reaction occurring within the initial components of acid–base cements [27, 39].

3.2.3.1 Zinc Phosphate Cements

Zinc phosphate is one of the conventional water-based dental cements introduced in 1879 for permanent cementing of indirect restorations. The formulation of these cements has been modified through the past century and it is currently used in dentistry considered as gold standard cements [35]. Zinc phosphate is a low-cost and non-adhesive cement and has been successfully used with metals, metal-ceramics, and ceramics. The cement reaches its maximal physical properties after 24 h and

Table 3.8 Some of the properties of commercial zinc phosphate cements [26, 27]

Property	Value
Setting time at 37 °C (minutes)	3.9–7.5
Film thickness (μm)	24–40
Compressive strength after 24 h (MPa)	70–131
Compressive modulus after 24 h (GPa)	11.9–13.5
Tensile strength after 24 h (MPa)	4.3–8.3
Solubility and disintegration after 24 h (%)	0.04–3.3
Bond to tooth	No
Excess removal	Easy
Fluoride release	No

presents great compressive strength. However, its tensile strength is rather low. On the other hand, the solubility of zinc phosphate is almost high, especially in acidic liquids. Due to the low pH of this cement, it is not suitable when there is a risk of pulpal irritation [26]. Due to the poor flexural strength of zinc phosphate, this cement is not suitable for luting ceramic and composite crowns [28]. The composition of this cement consists of zinc oxide powder, magnesium oxide powder (3–10%), phosphoric acid, water, and buffers [31]. The cements have been used for attaching all-metal, porcelain, and metal-ceramic indirect restorations as well as inlays, crowns, bridges, posts, orthodontic bands, and cavity liners in crowns and bridges because of their suitable mechanical retention [27]. Utilization of this cement as a temporary dental cement caries on but its inconvenience properties compared to resin-based cements including higher solubility, lower compressive and flexural strength, higher microleakage, dental pulp irritation, and lower bond strength has limited its utilization. The cementation process of zinc phosphate cements occurs rapidly and is highly exothermic making it necessary to control the reactions and dissipate the produced heat by mixing the precursors in a cool and dry container. After the completion of the setting reaction, the excess and unreacted phosphoric acid could dissolutely smear layer and plugs, drive into the dentinal tubules and cause pulpal irritation. The pH of cements varies from 2.14 after completion of mixing initial materials to 5.5 after 24 h of its implementation, which may cause post-operation pain [32, 35]. Table 3.8 represents some of the properties of commercial zinc phosphate cements.

3.2.3.2 Zinc Oxide-Eugenol (ZOE) and Noneugenol Cements

Zinc oxide-eugenol (ZOE) is one of the most popular water-based temporary dental cements produced by combining reactive zinc oxide with eugenol (4 allyl-2-methoxy phenol, a constituent of oil of cloves) and accelerators (acetic acid, zinc acetate, zinc alkanoates, zinc propionate and 2-ethoxybenzoic acid (EBA)). The hydrolysis reaction between eugenol (hydrogen ion provider) and zinc oxide is the basis of

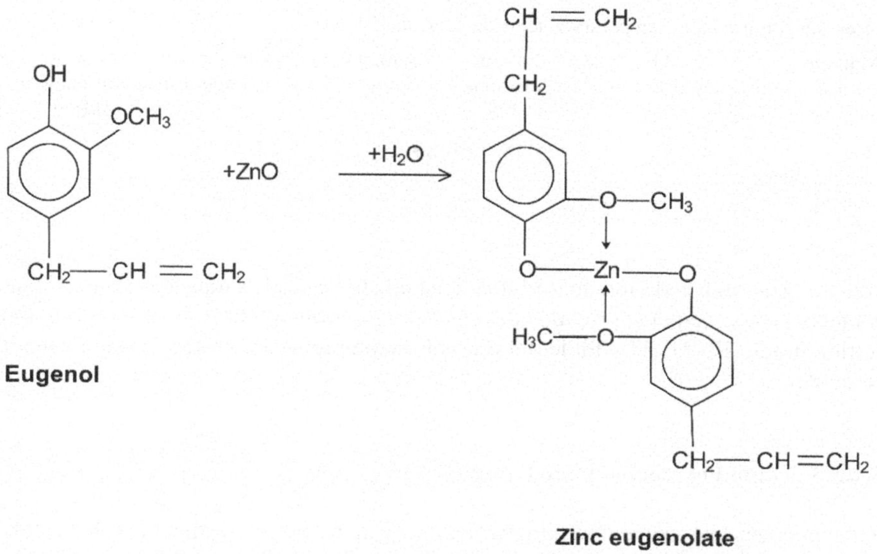

Fig. 3.24 The hydrolysis reaction between eugenol (hydrogen ion provider) and zinc oxide is the basis of the setting reaction [3, 27, 34]

the setting reaction in this cement leading to the formation of zinc eugenolate gel ($Zn(C_{10}H_{11}O_2)_2$), which is demonstrated as in the following Fig. 3.24 [3, 27, 34]:

ZOE is a favored dental cement owing to its antibacterial properties and convenience manipulation because of its moisture-sensitivity property, which causes to set rapidly in the oral environment besides its capability to have long working time (4–14 min) at room temperature. On the other hand, the compressive strength of this cement varies from 13 MPa to 38 MPa, which makes it useful to be used as liners and provisional cements. ZOE cements demonstrate setting shrinkage, which has high values of 0.86% and 0.32% for dry and wet conditions, respectively. In spite of the fact that the values of thermal expansion and shrinkage during the setting processes are high in these cements, their plastic behavior is attributed to convenient choices for sealing applications leading to minimize pulp inflammation. The ZOE cements have neutral pH and are more biocompatible compared to resin-modified glass ionomers and resin cements. On the other hand, eugenol releasement because of the hydrolysis of the salt matrix has pain-relieving properties. The introduction of ZOE cements into water or saliva leads to the breakdown of zinc eugenolate to zinc hydroxide and eugenol, which causes pulp inflammation. Zinc oxide-eugenol has a significant role in many applications of dentistry including sealing root canal, periodontal coating, cavity base, inelastic impression material, and temporary restorations (crown cementing and fillings). ZOE cements have been modified in the mid-1970 by introducing reinforced ZOE cements and orthoethoxybenzoic acid (EBA)-containing cements with better mechanical properties, which could be used as definitive cements. Contrary to ZOE cements, Eugenol-free or noneugenol (NE) cements

Table 3.9 Properties of temporary zinc oxide cements [33, 34]

Material	Working time (min)	Elastic modulus (GPa)	Setting time (min)	Compressive strength (MPa)	Tensile strength (MPa)
Zinc oxide-eugenol cements	>1.5	0.18–2.8	<7	2–5.5	0.4–1

are strong enough and more hydrolytically stable to be used as long-term provisional cements. These cements provide convenience retention and have long working and setting times [27, 31, 34]. Table 3.9 presents some properties of zinc oxide-eugenol cements.

3.2.3.3 Zinc Polycarboxylate Cements

Zinc polyacrylate cements (polycarboxylate), a member of the polyelectrolyte cements, introduced in 1968 by Smith was the first cement with chemical adhesion properties enabling it to make a bond with the calcium ions within the tooth structure. The aim of introducing this type of cements was to combine the convenient mechanical properties of zinc phosphate cements with the biocompatibility and bonding properties of ZOE cements. Zinc polycarboxylate cements set through an acid–base reaction between a powder (zinc oxide, magnesium oxide, aluminum oxide, and bismuth) and polyacrylic acid. The formation of chemical bonds while applying this water-based cements prevents microleakage of microorganisms, which makes it biocompatible and provides less irritation to the pulp compared to zinc phosphate cements. The major problematic issues of these cements, which have limited its utilization, are their severe solubility, poor mechanical properties, difficult manipulation, and their opaque esthetics, which makes it more favorable to be applied as a temporary cements rather than permanent one [26, 27, 31]. Table 3.10 illustrates some of the properties of zinc polycarboxylate.

Table 3.10 Properties of zinc polycarboxylate cements (Reprinted from Cambridge University Press, Wilson & Nicholson, Acid–base cements: their biomedical and industrial applications, no. 107, Copyright (2005), with permission from Springer) [27]

Property	Value
Working time at 23 °C	2–5 min
Setting time at 37 °C	3–12 min
Compressive strength (wet) after 24 h	48–80 MPa
Tensile strength (wet) after 24 h	4.8–15.5 MPa
Compressive modulus (wet) after 24 h	3.2–6.2 GPa
Adhesion to enamel after 24 h (tensile)	4.1–6.9 MPa
Adhesion to dentine after 24 h (tensile)	2.2–5.1 MPa

3.2.3.4 Glass Ionomer Cements (GICs)

Wilson and Kent first invented glass ionomer cements (GIC, aluminosilicate polyacrylic acid, or glass polyalkenoate cements) in 1969. These water-based cements are produced by mixing polycarboxylic acid, fine calcium-based fluoroaluminosilicate (FAS) glass, water, and tartaric acid [31, 40]. Glasses used in the mixture of these cements should be basic to be able to react with acid components. Different compositions are possible for these glasses and are commonly based on calcium and sodium compounds. G338 is one of these glasses, which is consisting of 24.9 wt% SiO_2, 14.2 wt% Al_2O_3, 4.6 wt% AlF_3, 12.8 wt% CaF_2, 19.2 wt% $NaAlF_6$, and 24.2 wt% $AlPO_4$. In some glasses, CaF_2 is replaced by SrO or SrF_2 in order to increase radiopacity. Tartaric acid and citric acid are chelating additives of GIC, which act as rate modifiers [41]. The setting reaction of glass ionomer cements results in the formation of salt by the reaction of polycarboxylic acid with the cations present in the glass. At the very first stage of the setting process, H^+ from the polyalkenoic acid attacks the glass and releases Al^{3+}, Ca^{2+} and Na^+ ions followed by the formation of calcium polyacrylate and aluminum polyacrylate. Later, with the releasement of F^- ions and silicic acid ($SiO_x(OH)_{4-2x}]_n$), a layer of salt gel (silica gel), which is stabilized by Al^{3+} ions forms on the ion-depleted surface of particles. By increasing the concentration of cations and hence the pH of the aqueous phase, an easier path brings about the cations to condense on the poly(acrylic acid) chains. Thus, the polyacrylate salts initially precipitate as a sol and then turn into a gel. This procedure continues and the cements harden until the last cation is bounded to the polyacrylate. Completion of setting process will never happen and the strength of cements rises until years. The set cements include calcium, aluminum, hydrogel, fluoroaluminum polyacrylates, and unreacted glass particles surrounded by a layer of silica gel [34, 42]. Figure 3.25 illustrates the setting reaction of glass ionomer cements.

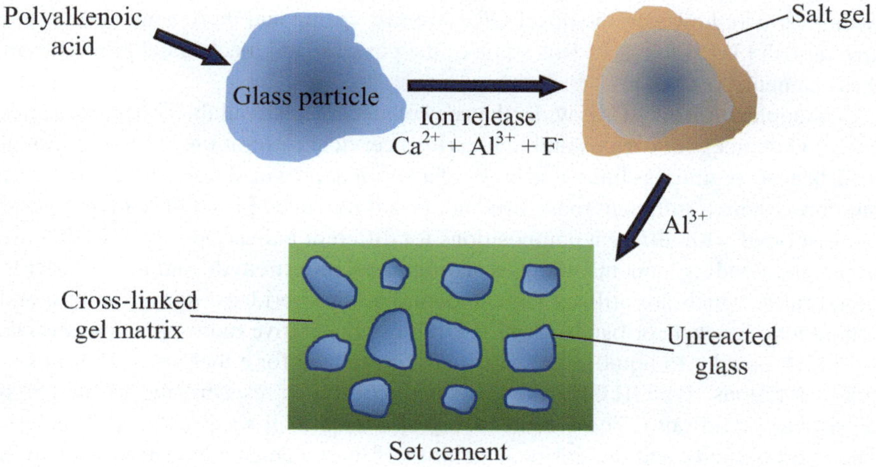

Fig. 3.25 The setting reaction of glass ionomer cements. Redrawn from von Fraunhofer et al. [34]

Table 3.11 Properties of glass ionomer cements [27, 31]

Property	Filling material	Luting agent
Film thickness (μm)	13.5	24–40
Working time (23 °C), minutes	1.3–3.8	2.3–5.75
Setting time (37 °C), minutes	2.75–4.7	4.5–6.25
Wet compressive strength (24 h), MPa	140–225	82–162
Wet compressive modulus (24 h), GPa	9–18	–
Wet flexural strength (24 h), MPa	8.9–39.1	4.1–15.5
Wet tensile strength, MPa	9.0–19.3	5.3–10.9

The presence of fluoroaluminosilicate glass filler in these cements avoids the dissolution of cements by phosphoric acid-containing zinc polycarboxylate and provides more translucency compared to polycarboxylate cements. The fluoride release of glass ionomer cements is a consequence of acid attack resulting in the entrance of fluoride ions into the cements matrix and giving the anti-cariogenic properties to the cements. Tartaric acid is stronger than polyacrylic acid and is added to the mixture to increase its working time by slowing down the formation of calcium polyacrylate. It also decreases the setting time of cements by facilitating the release of aluminum ions and formation of a complex between the polyacid and aluminum ions at lower pH compared to polyacrylic acid [3, 31]. GICs are stronger, less soluble, and more translucent than polycarboxylate cements [32]. Two significant characteristics of these cements are long-term fluoride release and their capability of providing molecular adhesion to tooth enamel and dentine as a result of phosphate and calcium ions chelation [31]. They are also capable of bonding to metals. Moreover, GICs present anti-cariogenic properties, similar-to-tooth thermal behavior, biocompatibility, and low cytotoxicity [40]. Due to the high solubility of conventional GICs during its first exposure with saliva, it was recommended to protect them temporarily with a varnish; however, the novel GICs are fast setting and there is no need to use any varnish [31]. Table 3.11 lists some of the physical and mechanical properties of these cements.

The application of GICs is wider than any other cements available. These materials are good choices for carious and non-carious cervical restorations, non or minimal load-bearing regions, as liners and bases of anterior approximal restorations, and core buildups where a sufficient tooth structure is still remained [44]. GICs are produced in three types with different compositions for different indications. Type I GICs are luting and bonding cements with fast setting, moderate strength, and low powder to liquid ratio, which are utilized for cementing crowns, bridges, inlays, onlays, and orthodontic brackets or bands. Type II GICs are restorative radio-opaque materials with high powder to liquid ratio, which are formulated for either anterior or posterior restorations. Type III GICs are liners and bases with, respectively, low and high powder to liquid ratio. These cements are also used as fissure sealants of molars. The hydrophilicity and dimensional stability of these cements have turned them in good choices over resin composites for this indication [41]. However, they are not recommended for restoring regions with high stress [31]. GICs present some drawbacks including poor esthetics, technique-sensitivity manipulation [44], low flexural strength, low fracture toughness, susceptibility to fail under large amounts of force because of the large Young's modulus, brittleness, high solubility, and low strength [32, 40]. Many attempts have been performed through the years in order to enhance the physical and mechanical properties of GICs [44]. Powder composition has been modified by inclusion of metallic particles, ceramic particles, nanoparticles, and phosphopeptides [45]. In this manner, metal powders have been frequently assessed resulting in the introduction of "cermets". These alloys containing cements are also called "metal-modified glass ionomer cements" (MMGICs), "packable GICs", "high viscosity GICs" or "high powder: liquid ratio GICs" [44]. Studies have shown that addition of SiC increases the risk of filler migration to vital organs resulting in the contraindication of this filler. On the other hand, addition of ceramic fillers such as ZrO_2 to some proportions increases the mechanical properties of the cements. Incorporation of metallic oxides and salts such as SrO, barium sulfate, and zinc sulfate were not successful due to the lack of strong chemical bonds between them and cements. The addition of hydroxyapatite also faced problematic issues due to the inconvenient mechanical properties of HA [40]. Orthophosphate-containing GIC, on the other hand, has shown remineralization owing to the formation of hydroxyapatite, fluorapatite, or strontium fluorapatite crystals. Moreover, nanoparticles have shown some advantages by providing nucleating sites and facilitating formation of fluorapatite and remineralization of the tooth. Until now there are a few studies regarding the modification of polymeric components. A few studies have assessed inclusion of amino acid-functionalized monomers, poly-N-vinylpyrrolidone, N-vinylpyrrolidone-modified acrylic acid copolymers, and star-shaped polymers. In this manner, inclusion of N-vinylpyrrolidone in these cements has provided better mechanical properties [44]. The latest studies have been carried out in order to improve the antibacterial release behavior of GICs as well as developing methods to control the antibacterial release of these cements. For the first case, the addition of chlorhexidine, epigallocatechin-3-gallate (EGCG), poly (quaternary ammonium salt) (PQAS), furanone, triclosan, and sodium fusidate in the composition of GIC has been assessed. Addition of propolis, strontium, salvadora persica (miswak), zinc sulfate, nanosilver, bioactive glass (BAG), and titanium dioxide nanoparticles to conventional GICs have also been studied in order to develop strategies for controlling antibacterial release in GICs [45].

3.2.3.5 Resin-Modified Glass Ionomer (RMGI) Cements

Considering several disadvantageous of water-based cements including high solubility, relatively low strength, and low adhesion, resin-modified glass ionomer (RMGI) cements were introduced in the 1980s to enhance the properties of dental cements. RMGIs also called hybrid ionomers (dual-cured hybrids) are a mixture of glass ionomers and resins set by acid–base and light-activated polymerization reactions. The water component of the glass ionomer cements could be substituted by water-hydroxymethyl methacrylate (water-HEMA), ethylene glycol dimethacrylate, glycidylmethacrylate, Bis-GMA besides polymerization initiator and activator systems [26]. The dual setting reaction of these cements could be represented as below:

(a) Acid–base reaction

Calcium aluminosilicate glass (base) + Polyacrylic acid = Calcium and aluminum polysalt hydrogel

(b) Photochemical polymerization

HEMA + photochemical initiator/activator = PolyHEMA matrix

The two above-mentioned reactions occur simultaneously while the acid–base reaction develops slower than the light-activated polymerization one [26]. The interaction between the polycarboxylate acid present in the cements and the hydroxyapatite present in the tooth structure leads to the formation of a strong ionic bond between the cements and the tooth. Furthermore, the micromechanical interlock introduced by the resin part to the tooth structure increases the strength of formed bonding. The bond strength is greater than that of glass ionomer cements; however, it is less than the bond formed between the standard resin-based cements and tooth, which is 6–10 times greater than the bonds formed by RMGIs [31]. The compressive and tensile strength of RMGI cements is more than glass ionomer cements and it is less soluble. Fluoride release and adhesive properties of RMGICs are similar to those of GICs. RMGI cements are widely used for cementing metal and metal-based crowns, bridges, inlays, onlays, and orthodontic brackets besides cementing zirconia-based ceramics, alumina-based ceramics, and lithium disilicate inlays and onlays. However, as a result of its high levels of fluid absorption and expansion, it is not a suitable cement to be used with low-strength all-ceramic crowns and veneers [27, 28]. RMGICs are produced in four types with different applications, which are type I (used as luting agent), type II (used as filling material), type III (used as liners and bases), and type IV (used as core buildups) [34].

3.2.3.6 Resin-Based Cements

Resin-based cements, the newest type of cements, are bonding cements with enhanced compressive, tensile, flexural, and adhesion properties. The first resin-based cements used in dentistry applications is methyl methacrylate-based introduced in 1952. The composition of resin-based cements consists of a resin (mostly dimethacrylate) and glass or silica fillers, which is similar to that of composite restoratives with lower filler concentrations in order to provide lower film thickness as well as improve the workability and setting time [31]. Adhesion of resin-based cements to the tooth structure is a consequence of micromechanical interlocking. Compared to other types of cements, resin-based cements are more resistant to dissolution, have improved mechanical properties (compressive, flexural, and tensile strength, toughness, wear resistance, and resilience), and can reduce microleakage. These cements could be used with all types of restorative materials including porcelains, metals, and composites. However, they have some drawbacks including polymer degradation, lack of fluoride release, probability of the existence of unreacted toxic monomers, lack of capability to perform strong or chemical bonds, high cost, and technique-sensitivity manipulation. Nowadays, these cements are produced in three categories for dentistry applications including etch-and-rinse resin cements (total-etch resin cements), self-etch resin cements, and self-adhesive resin cements (AR cements). The first two categories are also called conventional resin cements (CR cements) because of the need for pretreatments of the tooth in their application. The etch-and-rinse resin cements ordinarily show superior bond strengths to enamel, while the self-etch resin cements are often bonded to dentin with better adhesion strength. On contrary, self-adhesive resin cements show lower bond strengths compared to these two cements [27, 28, 31]. Table 3.12 shows the advantages, disadvantages, and applications of these three categories of resin cements.

Self-adhesive resin-based cements were introduced in 2002 presenting the capability of bonding to tooth tissue without following the three process steps of etching, drying, and priming, which are necessary in utilization of resin-based cements. These cements are the novel type of introduced cements with a wide range of popularity. Phosphoric acid-modified methacrylate monomers is the basic constituent of these cements, which provides self-adhesion properties as well as forming a cross-linked cements matrix, which supplies increased mechanical and dimensional stability. These cements go through a pH varying process in which their pH varies from around 2 at the first seconds of mixing to 7 after 24 h of their application. The acidity and hydrophilicity of cements in the first hours of mixing authorizes the self-adhesion of cements. At the pH values equal to 7, cements turn to be hydrophobic, which ensures stain-prevention of cements and also increases their long-term stability. However, polymer degradation over time remains an issue in utilizing these cements [31, 32, 34]. Table 3.13 lists various physical and mechanical properties of different cements used in dentistry applications.

3.3 Metals

3.3.1 Introduction

In the first chapter of this book, the history and variety of metallic-biomaterials are presented which cover the ones that are used in dentistry, too. The biocompatibility, a critical term in biomedical science, for metals group of dental materials are based on the structure, components, corrosion resistance, mechanical properties, and other critical factors. It is clear that corrosion can dramatically decrease mechanical properties

Table 3.12 Advantages, disadvantages, and applications of different types of resin cements [28]

Type of resin cements	Advantages	Disadvantages	Applications
Etch-and-rinse resin cements	High bond strengths to the enamel, sclerotic dentin, fluorosed enamel, etc. Good shade matching	Technique-sensitive, possibility of postoperative sensitivity if not used properly	1. Preparations where the enamel is predominant 2. Preparations on highly calcified tooth structures (fluorosis, sclerotic dentin, arrested dentin) 3. Enamel margins of inlays and onlays using the selective etch technique 4. Cementation of low-strength ceramics (feldspathic, porcelains, veneers) 5. Maryland bridges 6. High esthetic demand 7. Isolation is relatively easy (anterior teeth)
Self-etch resin cements	Good bond strength to the dentin, Ease of use (no need for a separate phosphoric acid etching step, rinsing and drying), Less risk of over-etching the dentin, Less risk of postoperative sensitivity	Relatively weaker bonds to the enamel and highly calcified tooth structures, May cause postoperative sensitivity if the adhesive is not air-dried, Most self-etch resin cements need refrigeration as acidic monomers degrade with heat, Less available shades compared to total-etch resin cements	1. Crowns and bridges with a lot of healthy dentin exposed 2. When retention is compromised: (a) Short crowns with less than 4 mm height (b) Crown tapers of more than 14° 3. Crowns and other fixed prostheses that have repeatedly come off 4. Inlays and onlays, especially those with deep cavities and few walls remaining 5. Preparations where dentin is predominant 6. Low esthetic demand 7. Isolation is difficult 8. Medium-strength restorations (lithium disilicate)

(continued)

Table 3.12 (continued)

Type of resin cements	Advantages	Disadvantages	Applications
Self-adhesive resin cements	Produces adequate bond to the dentin and enamel, less clinical steps are needed, less risk of contamination, Not as technique-sensitive as the etch-and-rinse cements	lower bond strengths compared to etch-and-rinse and self-etch resin cements, initially high bond strengths tend to decrease with time, when retention is the main issue and there is a lot of dentin exposed, the self-etch resin cements should take precedence over the self-adhesive cements as self-etch cements have higher bond strengths and are more durable	1. Use for areas of the mouth that are difficult to isolate such as the posterior segments 2. Used for most restoration types, regardless of the restoration material except veneers 3. For medium to high-strength materials (lithium disilicate, alumina, zirconia, PFM) 4. Indicated over conventional luting cements, when retention is compromised such as too tapered preparations, short crowns, and onlay preparations with little tooth structures and walls remaining and isolation and time constraints are a problem 5. Low esthetic demand

Table 3.13 Physical and mechanical properties of different cements used in dentistry applications [27, 31, 34, 38]

Cements	Compressive strength (MPa)	Tensile strength (MPa)	Elastic modulus (GPa)	Working time (min)	Setting time (min)	Solubility (weight % at 24 h)	Film thickness (μm)	Bond to tooth	Excess removal	Pulp irritation	Fluoride release
Zinc phosphate	96–133	3.1–4.5	13	1.5–5	5–9	<0.2	25.3	Non-adhesive	Easy	Moderate	No
Zinc polyacrylate	57–99	3.6–6.3	5–6	1.75–2.5	7–9	0.06	19	Molecular adhesion	Medium	Low	No
Zinc oxide-eugenol	13–52	1.2–2.8	0.18–2.8	4–14	7–9	0.1	25		Easy	Low	No
Glass ionomer	93–226	4.2–5.5	11.2	2.3–3.5	6–8	1	13.5	Molecular adhesion	Medium	High	Yes
Resin-modified glass ionomer	85–126	13–24	2.5–7.8	2–4	5–6	0.4–0.7	23.1	Molecular and micromechanical adhesion	Difficult	High	Yes
Compomer	194–200	36–40	17	3–10	3–7	Very low	>25	Molecular and micromechanical adhesion	Medium	High	Yes
Resin-based	180–265	34–37	4.5–9.8	0.5–5	1–15	0.05	29.6	Micromechanical	Very difficult	High	No
Self-adhesive resin-based	200–240	37–41	3–15	0.0–5	1–15	Very low	<25	Micromechanical	Very difficult	High	No

Table 3.14 Classification and application of the main types of metallic dental materials (Reprinted from Materials Science and Engineering: A, Upadhyay et al., Corrosion of alloys used in dentistry: A review, no. 3, Copyright (2006), with permission from Elsevier) [48]

Types of material	Classification	Application in dentistry
Amalgam	Low copper	Posterior restoration
	High copper	Posterior restoration
Noble alloys	Gold (non-heat-treatable)	Restoration of single teeth
	Gold (heat-treatable)	Fixed bridges
	Palladium–silver	Restoration of single teeth
Base alloys	Nickel	Restoration of single and missing teeth, partial denture frameworks
	Cobalt	Partial denture framework and implants
	Iron	Orthodontic appliances and endodontic instruments

of metal implant alloys such as strength and fatigue, which are attributed to the structural failure [46]. Metallic dental materials can be considered as two main groups. The first one is called base metals which are nickel–chromium, cobalt–chromium, and titanium-based alloys. The next branch is related to noble metals involving gold–platinum–palladium (Au–Pt–Pd), gold–palladium–silver (Au–Pd–Ag), gold–palladium (Au–Pd), and palladium–silver (Pd–Ag) [47]. Besides, other metallic implants with applications in dentistry are amalgam, gold, stainless steels, bulk metallic glasses (BMGs), and nickel–titanium (Ni–Ti). Three main types of metallic dental materials are presented in Table 3.14 with their classification and applications. In the following sections, the above-mentioned materials will be discussed.

3.3.2 *Base Metals*

Base metals do not indicate gold, silver, platinum, or palladium in the composition [49]. Generally, they can be divided into titanium and its alloys, cobalt–chromium, and nickel–chromium alloys.

Titanium (Ti) and Its Alloys. Titanium is known as a miracle metal in dental materials and the usage of that has been increased recently. This metal has been found by Klaproth for the Titans of Greek mythology. Its special mechanical properties caused it to be a widely used material in implant applications. Titanium has a high melting point in metals, with around 1700 °C. Commercially pure Titanium (Cp-Ti) is used for dental implants, surface coatings, crowns, partial, and complete dentures, which is presented in four grades that are different in the amount of oxygen and iron elements, with 0.18 wt% to 0.40 wt% (CP1-CP4) and 0.20–0.50 wt%, respectively, and having a directly impact on the physical and mechanical properties. Figure 3.26 shows the strength of Cp-Ti with oxygen contents variation [50].

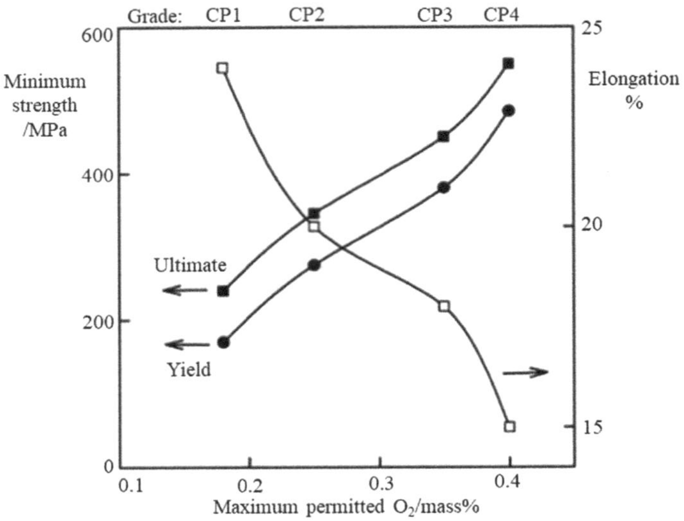

Fig. 3.26 The relation of strength and oxygen contents for Cp–Ti. (Reprinted from Regenerative Biomaterials, Liu et al., Binary titanium alloys as dental implant materials—a review, no. 316 Copyright (2017), with permission from Oxford university press) [50]

According to the The American Society for Testing and Materials (ASTM), there are three titanium alloys of Ti-6Al-4 V, Ti-6Al-4 V Extra Low Interstitial and Ti–Al–Nb, at which the Ti-6Al-4 V alpha-beta alloy has various applications in medicine, especially in dental implants. In Table 3.15, different physical properties of titanium and titanium alloys are presented [51]. However, it was reported that V contents show high cytotoxicity and Al content causes senile dementia and Alzheimer's problem. Different steps have been taken to tackle these problems such as new alloying elements were occupied for titanium-based alloys with satisfying consequences. For this, binary titanium-based alloys have been introduced like Ti–Nb, Ti–Ag, Ti–Au, Ti–Mn, Ti–Cr, Ti–Mo, Ti–Sn, Ti–Zr, Ti–Co, Ti–Pd, and Ti–Cu. Table 3.16 presents more information about the mentioned binary alloys. Additionally, cytocompatibility of some pure metals is ordered as Al > Ag > V>Mn > Cr > Zr > Nb > Mo > Cp–Ti [52]. In another study [53], it was concluded that the cytocompatibility of other pure metals is ordered as Cu > In > Ag > Cr > Sn > Au > Pd > Pt > Cp–Ti [50].

Titanium surface modifications are commercially available as dental materials but they are not much common in application. Surface composition and topography have a direct impact on events of the osseointegration that remains poorly understood. Surface topography (e.g. surface roughness) by establishing different methods can provide three levels of roughness on the scale of the feature which are macro-, micro-, and nano-sized topologies. Inorganic coatings like calcium affect the binding process of biologically active proteins from the peri-implant milieu as in its ionized form it adsorbs to the titanium dioxide surface and further to macromolecules with high affinity for Ca^{2+}. Better clinical results are available when calcium phosphate

Table 3.15 Physical properties of titanium and its alloys [51]

Physical terms	Statements
Biocompatibility	Excellent biocompatibilities due to the formation of TiO_2 layers and protective layers of semi- or nonconductive surface. Thin oxide layers around 4 nm provide inert, corrosion-resistant metal. Plus, the similarity of isolating effect of oxide layer by a dielectric constant (K) and water, prevent the detection of Ti as a foreign object by body environment
Osseointegration[a]	It is affected by surface roughness, composition, and degree of hydrophilicity. The total inert, great resistance to body fluid attack, great compatibility with bone growth leads Ti to be attached firmly to its place
Toxicity	Pure Ti and Ti–6Al–4 V alloy were applied as implant but the presence of V elements leads to toxicity in the human body. Ti–6Al–7Nb and Ti–5Al–2.5Fe have been introduced and developed as alternatives
Flexibility	The thermal coefficient of expansion and modulus of Ti is similar to that of human body which drop implant failure possibility. The stresses in working environment of implant cause failure and weakness of implant and surrounding bone. The flexibility of implant leads sharing less stress with surrounding bones that consequently save the bone from weakness
Non-magnetic properties	Cp–Ti and different titanium alloys are known as non-magnetic metals which account for preventing any susceptibility to outside interference. Additionally, this property provides the safety of magnetic resonance imaging examination

[a]Means that the apparent attachment or connection of osseous tissue to an inert, alloplastic substance without any intervening connective tissue

coatings like hydroxyapatite (HA) are applied that could advance the initial rate of osseointegration. Table 3.17 summarizes the disadvantageous and advantageous terms of Cp–Ti and Ti-6Al-4 V alloys based on surface modification techniques [55].

In dental titanium casting, commonly processes for fabricating are pressure-vacuum or centrifugal casting methods, at which the metal is melted via an electric plasma arc or inductive heating in a chamber with an inert gas atmosphere or in a vacuum environment. Subsequently, the molten metal is carried to the refractory mold by centrifugal or pressure-vacuum filling. Thanks to the new technologies and controlled fabricating environment, the consistency and efficient managing of the alfa-case layer (i.e. the coating of undesirable crust that is formed from incorporation of elements in investment and is attributed to the high reactivity and fragility) formation are achieved. However, ensuring the elimination of the alfa-case layer is far away which needs more study. Likewise, the formation of oxide layers on melted Cp–Ti surfaces is responsible for great chemical stability; however, this formed layer could weaken the reaction between the Cp–Ti and the boner ceramic components which affects negatively the Cp–Ti–ceramic union. Additionally, the oxide layer is attributed to porosity formation and metal-ceramic fracture. As a result, other processing methods should be developed for addressing this problem.

Dental crowns and bridge frameworks could be machined from solid metal stock by computer-aided machining (CAM) which, unfortunately, is not cost-effective and

Table 3.16 Some information about binary alloys of titanium in terms of additive effects, provided properties, etc. [50]

Binary Titanium Alloys type	Additive effects, properties, pros and cons
Ti–Zr	Zr is non-toxic, non-allergic, a transition metal, atomic number 40 and atomic weight 91.22 amu, melting point of 1857 °C and boiling points of 4409 °C Great resistance to corrosion due to the ZrO_2 layer formation which is osseo-integrated The continuous solid solution between titanium and zirconium Excellent properties of bioactivity, biocompatibility, and mechanics Ti–50 wt% Zr binary alloys proved higher hardness and tensile strength than Cp–Ti and Pure Zr
Ti–In	It is biocompatible that more addition of Into Ti improves mechanical properties, corrosion resistance, and biocompatibilities. For instance, Ti with 5–20 wt% In shows a similar corrosion resistance and higher oxidation resistance to Cp–Ti
Ti–Ag	The addition of Ag is to account for superior mechanical properties and corrosion resistance than Ti. Plus, toxicities of Ti–Ag are similar to Cp–Ti. Ti–5Ag and Ti–20Ag would be more appropriate for surface passivation film and cytotoxicity
Ti–Cu	It has a eutectoid structure at 7.0 mass % and intermediate phase of Ti_2Cu is presented in titanium-rich region. Some of the Ti–Cu alloys reveal higher mechanical properties because of solid solution strengthens the titanium and fine precipitation of intermetallic compounds. However, Ti with addition of Cu up to 10.0 wt% drop ductility but increase the yield and tensile strength. Plus, grindability will be improved with 5 and 10 wt% which is 2.6 times higher grinding rate than Cp–Ti
Ti–Au	This alloy has the eutectoid point and in the concentration of 15.3%, the intermetallic compound Ti_3Au forms. Ti–10Au illustrated the lowest value in galvanic corrosion current density test
Ti–Pd	Has adequate corrosion resistance and mechanical properties like hardness. Plus, this alloy proved good corrosion resistance against 0.2% NaF in artificial saliva
Ti–Nb	Excellent resistance to corrosion, Ti with more than 10% Nb show higher hardness, yield and tensile strengths than that of Cp–Ti. But dramatic lower tensile strength than Cp–Ti is reported
Ti–Mn	It was reported that Ti–Mn (8 and 12 Mn wt%) alloy with a plasma spark sintering preparation method provides high hardness and relative density in comparison with Ti alloys
Ti–Cr	Cr could provide Ti with a passive layer and increase corrosion resistance which under the condition of saline solution with F^-, 20 wt% of Cr give a greater corrosion resistance than Cp–Ti. The cast Cp–Ti has a hexagonal α phase and 5 wt% Cr addition will retain metastable β phase and 10 wt% and higher could retain equiaxed β phase
Ti–Mo	Highest bending strength could be reached by adding Mo contents with 10 wt% and 15 wt% of that because of fine BCC structure provide the lowest modulus among the β–Ti–Mo and also lower than Ti–6Al–4 V, Ti–6Al–7Nb, 316L stainless steel and Grade IV Cp–Ti.

(continued)

Table 3.16 (continued)

Binary Titanium Alloys type	Additive effects, properties, pros and cons
Ti–Sn	This binary alloy is non-toxic and non-allergic which the addition of Sn could strengthen Ti alloys. For instance, 1–30 wt% Sn contents result in HCP α structure and rising Vicker's hardness (H_V)
Ti–Co	Co is an acceptable biocompatible element which with regard to the Ti–Co diagram, at its eutectic or near eutectic composition, the significant melting point is provided and consequently, good castability and less metal-mold reactions happen
Ti–Ge	2 and 5 wt% provides non-toxicity, good mechanical performance, chemical corrosion resistance, and good processability. With 95.9 at % of Ti and the temperature of less than 882 °C, α-Ti is forming with reference to the Ti–Ge phase diagram. In open-air, Ge has leakage and the galvanic interaction with Ti causes corrosion. So, the usage of this element might not be viable so far.
Ti–Ga	Ga is a liquid metal with a dramatically low-melting-point if 29.77 °C leads to the lower temperature of processing. However, gallium was proved a galvanic interaction with Ti which the utilization of that may not be viable. 2 and 5 wt% provides non-toxicity, good mechanical performance, chemical corrosion resistance, and good processability

is limited. Electric discharge machining has been introduced as an alternative method that is using a fabricated graphite die to shape the metal to the desirable form via spark erosion. In the near future, the scientist should focus on a handy CAM method with controllable oxide layer formation on the Cp–Ti body [56].

The promising titanium implant is depended on the chemical and physical properties of the implant surfaces which directly affect the bone response. A commonly used manner for satisfying this demand is anodic oxidation surface treatment which brings corrosion resistance, aspect improvement, joining with polymers, and so on. This method creates a porosity which improves the anchorage of the implants to the bone and increases the possibility of antibiotics flowing out around Ti implants. Likewise, plasma-induced graft polymerization is another method for introducing functional surfaces of implants. As an example, Ti/NH_2 modified samples with immobilized HA (hyaluronic acid) have been anodized in order to enhance osteoblast-like cell attachment and proliferation which is a promising option for implant application [57].

Cobalt–Chromium (Co–Cr). One of the most popular base metal alloys is the Co–Cr alloy because of excellent corrosion resistance. The existence of chromium is to account for strengthening and passivation for corrosion resistance improvement. The composition and some mechanical properties of this alloy are presented in Tables 3.18 and 3.19, respectively. This group could be divided into two main groups which are ruthenium and ruthenium-free. This alloy could be used instead of nickel-containing alloys which have an allergic consequence for some patients. A new class of Co–Cr alloys is involving Au, Pt, Ru, or In, Ga, Mn, and W for achieving effective bonding

Table 3.17 The advantages and disadvantages of Cp–Ti and Ti–6Al–4 V alloys according to the surface modification techniques (Reprinted from Journal of Materials Science, Duraccio et al., Biomaterials for dental implants: current and future trends, no. 4783, Copyright (2015), with permission from Springer) [55]

Surface modification tech.	Advantages	Disadvantages
Roughening		
Plasma spraying	Mechanical anchorage and fixation to bone are favored	Titanium wear particles in the bone
Grit blasting	Mechanical anchorage and fixation to bone are favored, the high survival rate	The residue of blasting material interfered with osseointegration
Etching	Protein adsorption, osteoblastic cell adhesion, and rate of bone tissue healing in the peri-implant region	Reduction in mechanical properties of Titanium
Anodization	Protein adsorption, osteoblastic cell adhesion, and rate of bone tissue healing in the peri-implant region	Process rather complex
Grit-blasted and acid-treated surface (SLA)	A high rate of bone formation in early stages of peri-implant bone regeneration and enhanced bone-implant contact in areas of surfaces previously not covered by bone	Possible surface contamination with hydrocarbons
2-step treatment	Accelerated bone tissue regeneration and increased mechanical retention	Multistep process and high temperature
Coating		
Plasma-sprayed HA[a] coating	High integration rate, fast bone attachment, direct bone bonding, a high initial rate of osseointegration	Coating delamination, Controversies regarding long-term Prognosis
Ion implantation	High percentage BIC values Process extremely controllable. Possibility to have ultra-high-purity layers	An expensive process, no clinical performance

[a]hydroxyapatite (HA)

characteristics to the ceramic substrate which are not much effective for improving corrosion resistance and in some cases, they are reducing that. For instance, adding Au, Pt, and In elements cause a heterogeneous microstructure with spherical Au–In and Au–In–Pt precipitates which could be account for the degeneration of the corrosion resistance [47, 58].

Nickel–Chromium (Ni–Cr). The composition of the Ni–Cr alloy is presented in Table 3.20 at which the percentage of that is changing between 62 and 77 which directly affects the physical and corrosion resistance properties. In some cases, Al and Ti are applied to provide strengthening precipitation of Ni_3Al and Ti_3Al, and also Fe, W, and V are added for solid solution hardening. Likewise, Mo-rich Ni–Cr shows better mechanical and thermal coefficient properties. The mechanical properties of this alloy are mentioned in Table 3.21 [47].

In general, the casting temperature range of Co–Cr or Ni–Cr alloys is around 1300–1450 °C which is relatively higher than other dental materials like gold with casting temperature more than 950 °C. As a consequence, this property puts a strain on the casting process at which an acetylene-oxygen flame or an electric induction furnace is needed. The condition of the process is important at which too much oxygen may lead to oxidation, on the other hand, an excess of acetylene causes embrittlement due to the increase of metal carbide content. Likewise, the accessories of the process like investment molds should endure the structure and application under the high temperature. For this, silica-bonded and phosphate-bonded are suggested at which the latter is more commonly used. Another factor that limits the production of base metals specially Co–Cr alloys is being time-consuming of preparation due to the trimming and polishing that consequently increase the final product price [59].

3.3.3 Noble Metals

Noble alloys are popular due to the chemically inert character in body fluid condition and in the oral cavity which causes them to not be susceptible to corrosion attack and they are introduced in the next subsections [46]. Tables 3.22 and 3.23 present the chemical compositions and mechanical properties of noble metal alloys.

Gold–Platinum–Palladium (Au–Pt–Pd). This alloy contains Au with 75–88 wt% and less than 8 and 11 wt% for Pt and Pd, respectively. It was firstly applied for metal-ceramic operative materials but after the arrival of different cost-effective alloys with better mechanical properties and sag resistance, the usage of that has been dropped. So, the application of Au–Pt–Pd alloys in crowns should be limited [47].

Gold–Palladium–Silver (Au–Pd–Ag). Due to the limitation of Au–Pt–Pd alloys, the alloys of Au–Pd–Ag have been introduced which is divided into high or low silver portion. However, the silver content may cause color effects on porcelain [47].

Table 3.18 Composition of Co–Cr alloys (wt%), (Reprinted from Journal of Prosthodontics: Implant, Esthetic and Reconstructive Dentistry, Roberts et al., Metal-ceramic alloys in dentistry: a review, no. 190, Copyright (2009), with permission from John Wiley & Sons) [47]

Co	Cr	Mo	Al	Fe	Ga	Nb	W	B	Ru
53–68	25–34	0–4	0–2	0–1	0–3	0–3	0–5	0–1	0–6

Table 3.19 Properties if Co–Cr alloys (wt%), (Reprinted from Journal of Prosthodontics: Implant, Esthetic and Reconstructive Dentistry, Roberts et al., Metal-ceramic alloys in dentistry: a review, no. 190, Copyright (2009), with permission from John Wiley & Sons) [47]

Ultimate tensile strength (MPa)	0.2% yield strength (MPa)	Elastic modulus (GPa)	Elongation (%)	Diamond pyramid hardness (Kg/mm^2)	Casting temperature (°C)
520–820	460–640	145–220	6–15	330–465	1350–1450

Table 3.20 Composition of Ni–Cr alloys (wt%), (Reprinted from Journal of Prosthodontics: Implant, Esthetic and Reconstructive Dentistry, Roberts et al., Metal-ceramic alloys in dentistry: a review, no. 190, Copyright (2009), with permission from John Wiley & Sons) [47]

Ni	Cr	Mo	Al	Fe	Be	Ga	Mn
62–77	11–22	4–14	0–4	0–1	0–2	0–2	0–1

Table 3.21 Mechanical properties of Ni–Cr alloys (wt%), (Reprinted from Journal of Prosthodontics: Implant, Esthetic and Reconstructive Dentistry, Roberts et al., Metal-ceramic alloys in dentistry: a review, no. 190, Copyright (2009), with permission from John Wiley & Sons) [47]

Ultimate tensile strength (MPa)	0.2% yield strength (MPa)	Elastic modulus (GPa)	Elongation (%)	Diamond pyramid hardness (kg/mm^2)	Casting temperature (°C)
400–1000	255–730	150–210	8–20	210–380	1300–1450

Gold–Palladium (Au–Pd). It was firstly introduced in 1977 with white gold color for copping the demands for silver-rich alloys which were porcelain discoloration and high thermal expansion coefficient. However, the main drawback of this alloy is the very low compatibility of degree of thermal expansion with some high expansion porcelains. For this disadvantageous, some silver content (<5%) was added and as a consequence, the discoloration of the porcelain was terminated, castability is improved, and the thermal expansion coefficient is raised [47].

Table 3.22 Composition of noble alloys (wt%), (Reprinted from Journal of Prosthodontics: Implant, Esthetic and Reconstructive Dentistry, Roberts et al., Metal-ceramic alloys in dentistry: a review, no. 189, Copyright (2009), with permission from John Wiley & Sons) [47]

Type	Au	Pt	Pd	Ag	Cu	Sn	Ga	In	Other
Au–Pt-Pd	75–88	≤8	≤11	≤5	–	2–5	–	<1	Fe, Re
Au–Pd–Ag	39–77	–	25–35	12–22	–	3–7	–	1.5	Fe, Ru, Re
Au–Pd	44–55	–	35–45	–	–	8–12	≤5	8–12	Ru, Re
Pd–Ag	–	–	50–60	28–40	–	4–8	–	1–5	Ru

Palladium–Silver (Pd–Ag). The first introduction of this alloy was in 1974 for being an alternative to high-cost gold-based alloys. Generally, there are two main groups of Pd–Ag alloys based on the composition. One of them is 55–60% palladium-rich and 28–30% silver-rich. The second group has 50–55% palladium and 35–40% silver, tin, and other trace materials, with little or no indium. One of the most attractive properties of Pd–Ag alloys is the excellent sag resistance due to the high elastic modulus. Additionally, it has a good porcelain bond strength.

However, Pd–Ag alloys showed more porcelain discoloration than Au–Pd–Ag alloys [47].

3.3.4 Amalgam

Generally, dental amalgams are formed by mixing 45–55% Hg and a powder mixture of Ag, Sn, and Cu (in some cases, smaller amounts of Zn, Pd, In, or Se are added) by applying a mechanical vibration process, which results in a putty-like material that is account for easily filling and manipulation. Dental amalgam is currently widely used as a dental restorative material and has been used over 150 years after firstly introduced by Chinese and in 1528 it was proposed as a dental filling by a European physician in Germany. The spot parts of the amalgam history date back to a silver–mercury amalgam that was made in France. A Silver paste filling material was fabricated by mixing the silver coins with mercury. Between around 1895–1908 years, the amalgam was globally accepted with the development of state-of-the-art alloy with the composition of 68.5% silver, 25.5% tin, 5% gold, and 1% zinc and the first commercially available amalgam rich in silver and gold as an alternative to the copper has happened. By introducing a new atomization process for dental amalgams manufacturing, extreme improvement in the quality has occurred. By addition of copper elements, with high portion, the durability of amalgam has been increased which causes 12 years serving and consequently, the post-processing of the polishing is not required after placement [60].

Table 3.23 Mechanical properties of some noble metal alloys (Reprinted from Journal of Prosthodontics: Implant, Esthetic and Reconstructive Dentistry, Roberts et al., Metal-ceramic alloys in dentistry: a review, no. 189, Copyright (2009), with permission from John Wiley & Sons) [47]

Type	Ultimate tensile strength (MPa)	0.2% yield strength (MPa)	Elastic modulus (GPa)	Elongation (%)	Diamond pyramid hardness (kg/mm^2)	Casting temperature (°C)
Au–Pt-Pd	480–500	400–420	81–96	3–10	175–180	1150
Au–Pd–Ag	650–680	475–525	100–113	8–18	210–230	1320–1350
Au–Pd	700–730	550–575	100–117	8–16	210–230	1320–1330
Pd–Ag	550–730	400–525	95–117	10–14	185–235	1310–1350

There is no doubt to say that the oral environment is potential for corrosion attacks due to the wide pH range during the mastication of different foods. So, any restorative materials would be susceptible to corrosion problems. Amalgam has a high corrosion resistance in most of all oral atmospheres. But in some cases when the level of fluoride content increases, the corrosion resistance is negatively affected. Amalgam often becomes tarnished (i.e. an initial sign of corrosion at which a thin black layer grows across the surface of the alloys containing microorganisms and mucin) and surface discoloration happened due to the oxides and sulfides formations. The corrosion of that could trigger by using galvanic couples and as a result, ion releasing is occurred [61]. The ions releasing especially mercury contents face the body with different toxic problems such as neurotoxicity, kidney dysfunction, etc. The techniques for inspection of initial corrosion are limited because of current clinical diagnostic tools and under surface corrosion of restoration materials. Conventional silver amalgam materials involving phases that harshen corrosion attacks to the γ_2 phase (Sn_7Hg). As a consequence, the formation of the tin oxychloride is account for toxic mercury release and formation of the γ_1 and γ_2 from unreacted γ phase. In high copper amalgam alloys, the ή (Cu_6Sn_5) phase is vulnerable to the corrosion which does not release mercury into the body. However, due to the shifts of E_{corr} to more positive values, high Hg injection might still happen. Electrochemical testing on pure phases proved that the Ag_2Hg_3 phases could dramatically shield against corrosion attacks followed by Ag_3Sn, Ag_3Cu_2, Cu_3Sn, Cu_6Sn_5, and $Sn_{7-8}Hg$. Moreover, the compounds of SnO, SnO_2, $Sn_4(OH)_6Cl_2$, Cu_2O, $CuCl_{2.3}Cu(OH)_2$, $CuCl$, $CuSCN$, and $AgSCN$ have been observed in patients [62]. In spite of the controversy about amalgam usage, the World Dental Federation claims that dental amalgam has a well-documented background of harmless results and effectiveness during application [63].

3.3.5 *Gold*

It is well known as a chemically inert material, and ancients until 1700 s used to apply it for wires to hold replacement teeth in place. The root of the lost tooth has been separated and subsequently drilled to establish the gold wire which has been wrapped around the adjacent tooth as a supporter. With the arrival of porcelain teeth in 1780 s, swaged gold foil was made in order to enhance the structural integrity. Till 1930, cast gold alloys have been used to fasten the denture to the remaining teeth for partial dentures and were denied after the introduction of the cobalt–chromium alloy. The crystal structure of gold is a face-centered cubic (FCC) which leads to be compatible with different ductile metallic elements like aluminum, silver, nickel, platinum, and copper. In the contamination-free form of gold, all three of high symmetry planes undergo reconstruction but, all in all, it is far away to say that all three of them could reconstruct at which (110) and (100) surfaces (with surface energy of 1.28 and 1.60 J.m-2, respectively) may reconstruct. However, the FCC structure causes this metal to be ductile. Gold could be applied as a dental material purely or in alloying with other noble metals. Although the pure form of application is limited due to the poor mechanical properties, it is very soft (HV 25), with a dramatically low 0.2% proof stress (30 MPa) and a big elongation (45%) which are attributed to be easily cold worked. So, it is just suitable for very small cavities. Electroformed inlays and onlays are appropriate for cemented into cavities after covering with porcelain [64].

Cast gold alloy inlays and onlays applying into cavities with cementing investment method is more common at which Table 3.24 indicates related standards and compositions.

High sag resistance and acceptable coefficients of thermal expansion in comparison with porcelains trigger the usage of porcelain-fused-to-metal restorations for fixed crowns and long-span bridges with porcelain veneers area. Table 3.25 provides information about gold alloys properties [65].

3.3.6 Stainless Steels (SS)

The history of the stainless steels dates back to 1912 when they were firstly introduced by Brearly of Sheffield for the rifling of gun barrels [66]. The first application of that in orthodontics was in the early 1930s at which gold alloys played a central role in wires fabrication [67]. Nowadays, the annual application of stainless steel has been increased due to the different facts especially being cost-effective. The main composition of the stainless steel is Fe alloys with the addition of Cr around 10–20 wt% at which it accounts for corrosion resistance (200 times higher than common carbon steel) by surface oxidation and protection process. For the enhancement of other properties like mechanical properties, different chemical elements such as Ni, Mo, and Ti are added. Generally, stainless steels could be divided into austenitic, ferritic, martensitic, duplex, and precipitation hardening alloys (Table 3.26). Moreover, the mechanical properties and corrosion resistance of main stainless steels are presented in Figs. 3.27 and 3.28 [68].

A wide variety of stainless steel is occupied for fabrication of different brackets like AISI 304L, 316L, 17–4PH stainless steel, and 2205 alloy. Likewise, mini implants besides titanium alloys are made by stainless steels and orthodontic wires which are mostly fabricated by stainless steels from types of 302 and 304 18/8 austenitic alloys [69] (Fig. 3.28).

3.3.7 Bulk Metallic Glasses (BMGs)

In 1960, metallic glasses have been arrived with the production of the Au75Si25 metallic glass alloy. These amorphous alloys or liquid metals showed excellent toughness and corrosion resistance which are critical in biomedical applications. By the introduction of bulk metallic glasses (BMGs) with really lower critical cooling rates around <100 K/s, the limitation of the product's dimensions has been solved.

Table 3.24 Dental gold alloys with needed standards, usage, and composition (Reprinted from Gold Bulletin, Knosp et al., Gold in dentistry: alloys, uses and performance, no. 94, Copyright (2003), with permission from Springer, CC BY 4.0) [65]

Alloy number	Gold element % (m/m)	Another elements % (m/m)	ISO EN Standards	Usage
1	99.9–99.99	–	–	A, M
2	97.9–98.3	1.7 Ti, Ir, Rh, Nb	9693	C, D, E
3	75–90	10–20 PGMs, In Sn, Fe, Re, Ag, Cu, Zn, Ta, Ti, Mn	9693	C, E
4	60–75	roughly 10 PGMs, 10–25 Ag, Cu, In, Sn, Zn	1562 9693	B, C, D
5	65–75	5–10 PGMs, 5–20 Ag, Cu, Zn	1562	B, D, E, F, G, H, L
6	60	40 PGMs	–	H, I, N
7	40–60	5–10 Pd, 10–30 Ag, Pt, Cu, Zn, Ir	8891	B, D, L
8	40–60	20–40 Pd, 0–20 Ag, In, Sn, Cu, Ga, Ir, Re, Ru	9693	C
9	Roughly 15	roughly 50 Pd, roughly 20 Ag, In, Sn, Ga, Ir, Re, Ru	9693	C
10	2–12	30–60 Ag, 20–45 PGMs, Cu, Zn, Ir,	8891	B, F, L
11	0.1–2	50–60 Pd, 25–40 Ag, In, Sn, Ga, Ir, Re, Ru	9693	C
12	0.1–5	70–80 Pd, Sn, Cu, Co, Ga, Ir, Re, Ru, Pt	9693	C
13	10–80	0–20 PGMs, 0–70 Ag, Cu, Sn, In, Zn, Co, Ni, Mn	9333	K

Remarks (A) Electroforming, (B) Inlays, Onlays, Fixed Crown- and Bridgework as Full Cast and for Polymer Veneers, (C) Fixed Crown- and Bridgework for Porcelain Veneers, (D) Telescopic Crowns and Milling Work for Removable Dentures, (E) CAD/CAM, (F) Orthodontic Wire, (G) Clasp Wire, (H) Prefabricated Attachments, (I) Root Canal Posts, (K) Solders, (L) Partial Denture Framework, Cast Attachments, Plates, Connecting Bars, Saddles and Splints, (M) Gold Foil for Direct Fillings, (N) Cast-On Technique
PGMs = Platinum Group Metals

The combination of the amorphous structure and metallic materials properties leads to amazing mechanical properties like high strength and elasticity, which makes BMGs more attractive. Figure 3.29 illustrates the connection of bioglass, bio-metallic

Table 3.25 Gold alloys properties (Reprinted from Gold Bulletin, Knosp et al., Gold in dentistry: alloys, uses and performance, no. 96, Copyright (2003), with permission from Springer, CC BY 4.0) [65]

Alloy number from Table 3.11	Composition % (m/m)										Color	Usage-Based in Table 3.11	Hardness as-cast or Hardened Resp (HV)	0.2% proof stress as-cast or Hardened[a] resp (MPa)	Elongation as-cast or hardened[a] Resp (%)	Melting Range (°C)	Coefficient of thermal expansion (25–500 °C) (µm/m K)
	Au	Pt	Pd	Ag	Cu	In	Sn	Zn	Ir	Other							
2	98.2	–	–	–	–	–	–	–	0.1	1.7Ti	Yellow	C, D, E	200	415	6	1060–1105	14.5
3	84	8.2	4.7	0.5	0.1	2.4	–	–	0.1	–	Pale Yellow	C	190	630	7	1085–1195	14.3
	77.7	19.5	–	–	–	–	–	2	0.1	0.7Ta	Pale Yellow	C, E	220	530	17	1050–1165	14.0
4	73.8	9	–	9.2	4.4	1.5	–	2	0.1	–	Yellow	B, C, D	200	460	5	900–950	16.7
5	87.4	–	1	11.5	–	–	–	–	0.1	–	Deep Yellow	B, Type 1	55	80	45	1030–1080	–
	77	1	–	13	8.5	0.2	–	0.2	0.1	–	Deep Yellow	B, Type 2	120	210	54	900–940	–
	78	4.9	–	11	4	–	–	2	0.1	–	Deep Yellow	B, Type 3	195	380	18	920–970	–
	70	3.9	2	13	9.5	–	–	1.5	0.1	–	Yellow	B, D, E, F, G, H, L, Type 4	250	600	12	895–935	–
6	60	24.9	15	–	–	–	–	–	0.1	–	White	H, I, N	230[a]	730[a]	10[a]	1320–1460	–
7	50.1	–	10	29.8	–	8	–	2	0.1	–	Pale Yellow	B, Type 3	200	440	9	880–965	–
	57	0.5	4	24.9	13	–	–	0.5	0.1	–	Yellow	B, D, L, Type 4	260	700	10	870–905	–
8	51.5	–	38.4	–	–	8.5	–	–	–	1.5 Ga, 0.1 Ru	White	C	225	575	19	1170–1310	13.7

(continued)

Table 3.25 (continued)

Alloy number from Table 3.11	Composition % (m/m)										Color	Usage-Based in Table 3.11	Hardness as-cast or Hardened Resp (HV)	0.2% proof stress as-cast or Hardened[a] resp (MPa)	Elongation as-cast or hardened[a] Resp (%)	Melting Range (°C)	Coefficient of thermal expansion (25–500 °C) (μm/m K)
9	15	0.1	52.3	20	–	6	5.5	–	–	1 Ga; 0.1 Re	White	C	265	620	6	1145–1265	14.1
10	10	–	20	54.2	15	–	–	0.8	–	–	White	B, Type4	200	450	13	900–970	–
	10	10	35	30	15	–	–	–	–	–	White	F, G	315[a]	850[a]	15[a]	1060–1110	–
11	2	–	57.8	28	–	4	6	2	–	0.2Ru	White	C	260	650	8	1175–1275	14.9
12	1	1	79.7	–	5	–	6.5	–	–	6GA; 0.8 Ru	White	C	260	575	30	1155–1290	13.8
13	76	2.9	–	10	6	–	–	5	0.1	–	Yellow	K	–	–	–	850–880	–
	79.8	–	3	15.5	0.1	0.3	1	0.1	0.2	–	Yellow	K	–	–	–	1025–1055	–
	79	–	–	1.5	0.7	–	–	2	–	16 Ni; 0.8Mn	Pale Yellow	K	–	–	–	850–920	–
	10.2	–	8.1	62.8	11.1	4.9	2.9	–	–	–	White .	K	–	–	–	855–885	–

Remarks Type 1: low-strength—for castings subject to very slight stress such as inlays
Type 2: medium-strength—for castings subject to moderate stress such as inlays and onlays
Type 3: high-strength—for castings subject to very high stress such as onlays, thin cast backings, pontics, full crowns and saddles
Type 4: extra-high-strength—for castings subject to very high stress and thin cross-section such as saddles, bars, clasps, thimbles, unit castings and partial frameworks

Table 3.26 Different alloys of stainless steel with composition and mechanical properties [68]

Alloy system	Definition
Austenitic	The biggest family of stainless steels wit application in food processing Equipment, utensils for the kitchen, and medical equipment. They are non-magnetic and not heat-treatable. They have three branches: Cr–Ni (300 series), Mn–Cr-Ni–N (200 series) and specialty alloys
Ferritic	Good ductility, formability, and poor high-temperature behavior are the main characteristics. Plus, these are magnetic and non-heat-treatable with cost-effective price than other stainless steels. The main composition is very low Ni content, Cr content changing from 12 to 17 wt%, C content less than 0.1 wt%. Other additive elements like Mo, Al, Ti are optional
Martensitic	Hardenable steel in which carbon content has a major impact on formability. These are magnetic and heat-treatable with application in dental and surgical equipment. The composition of that are Cr with the range of 11–17 wt%, Ni less than 0.4 wt% and high C content with 1.2 wt%
Duplex	High yield strength and stress corrosion cracking resistance in a chloride-rich atmosphere which is overweight that of austenitic stainless steels. Due to the presence of two-phase of Ferrite and Austenitic, high strength with excellent ductility is reachable. The composition of duplex group is Cr with 22–25 wt% and Ni with 5 wt% by the addition of Mo and N contents
Precipitation Hardening	A Cr–Ni stainless steels involving Al, Cu, and Ti as an alloying agent which provide more hardness by solution and aging heat treatment

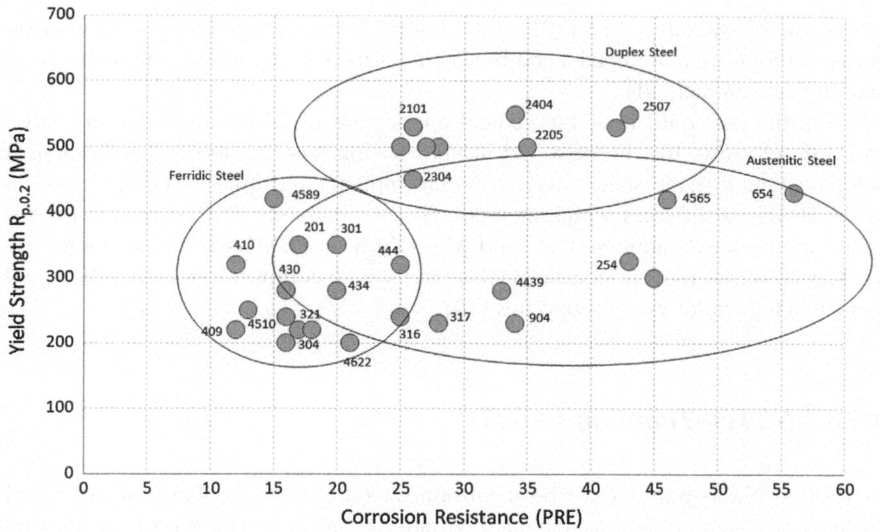

Fig. 3.27 The graph of corrosion resistance in relation to yield strength of stainless steel. (Reprinted from Construction and Building Materials, Corradi et al., A review of the use of stainless steel for masonry repair and reinforcement, no. 338, Copyright (2018), with permission from Elsevier) [68]

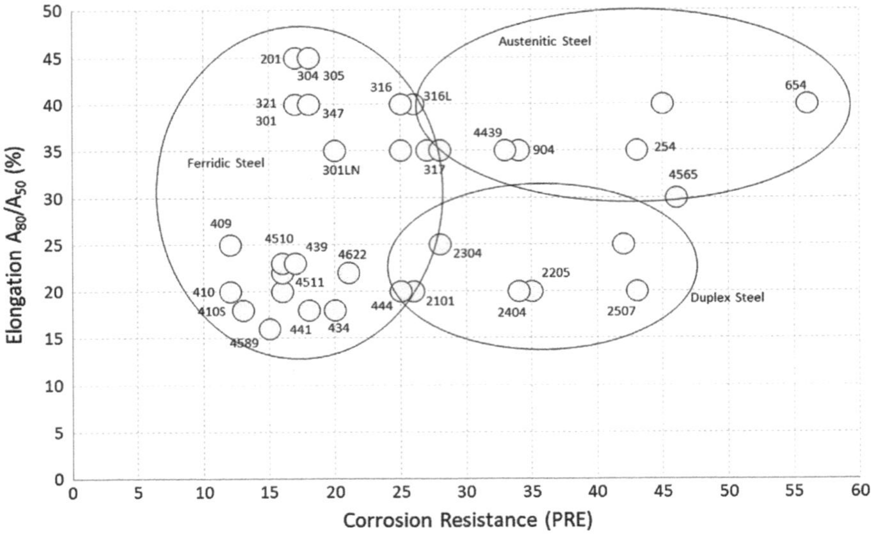

Fig. 3.28 The graph of corrosion resistance in relation to elongation of stainless steel. (Reprinted from Construction and Building Materials, Corradi et al., A review of the use of stainless steel for masonry repair and reinforcement, no. 338, Copyright (2018), with permission from Elsevier) 68]

alloys, and bio-metallic glass [70]. Additionally, Table 3.27 represents a comparison between bioglasses, biometals, and BMGs in terms of composition, microstructure, and preparation methods.

From the past until now, biomedical application of BMGs has been increased and different tests like in vitro and in vivo testing were established for practical behavior which is the same way for dental implants and fillers. The composition of the BMGs is different which includes Ti-, Zr-, Fe-based BMGs that are the so-called amorphous stainless steels and Mg-, Zn-based BMGs are other members. Table 3.28 shows most important BMGs family with composition, preparation, and some mechanical/physical properties [70].

3.3.8 Nickel–Titanium (Ni–Ti)

In 1960 s, Ni–Ti alloys have been introduced for a space program at the Naval Ordnance Laboratory and were named Nitinol (i.e. *ni* for nickel, *ti* for titanium, and *nol* named after Naval Ordnance Laboratory) that are super-elastic intermetallic alloys. This alloy provides a shape memory effect, which can be controlled by programmed, specific heat treatment. About the structure, the body-centered cubic (BCC) lattice with austenite phase at a temperature higher than 100 °C is steady. Cooling through the critical transformation range (TTR) leads to an extreme altering

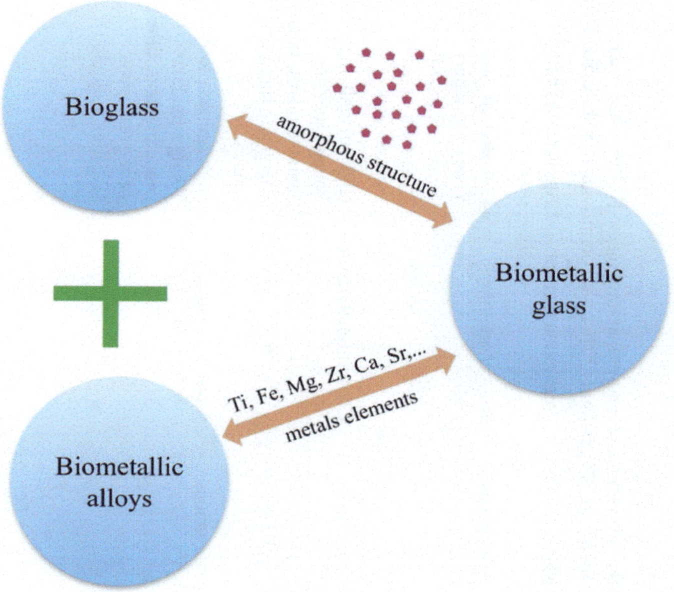

Fig. 3.29 The relationship of bioglasses, bio-metallic alloys, and bio-metallic glasses. Redrawn from Li and Zheng [70]

in stiffness, yield strength, and electric resistivity due to the electron bonding changes, and consequently, martensitic transformation, which is directly related to the starting and finishing temperature of cooling happened, and subsequently the characteristic of shape memory alloy (SMA) is given. 55-Nitinol represents alloys with roughly 56 wt% nickels and 44 wt% titanium at which in some cases around <2% of nickel can be substituted by cobalt. These alloys are capable of altering their type of atomic bonding, which accounts for precious and important changes in mechanical properties and lattice atomic arrangements. The two important phases of Nitinol between the temperature of 0 and 100 are austenite and martensite which are, respectively, high temperature and low temperature [71]. The physical and mechanical properties with preparation methods of the Nitinol alloys are given in Table 3.29.

The first introduction for biomedical application of Nitinol as an implant material was around 1970 s. This shape memory alloy is applied in orthodontics for brackets and wires connecting the teeth at which by raising the temperature to ambient body temperature, the Nitinol causes a constant force to move the teeth by contracting back to the original shape. Additionally, Nitinol files could be used as a cleaner and shape forming for root canal procedures [51].

Table 3.27 Comparison between bioglasses, biometals, and BMGs in terms of composition, microstructure, and preparation methods (Reprinted from Acta Biomaterialia, Li and Zheng, Recent advances in bulk metallic glasses for biomedical applications, no. 3, Copyright (2016), with permission from Elsevier) [70]

Materials	Composition	Classification	Microstructure	Preparation manner	Application and properties
Bioglasses	Na_2O–CaO–Si_2O–P_2O_5 oxides	Biodegradable: 45S5, phosphate-based glasses, borate-based glasses Non-biodegradable: 60S3.8, 65S	Amorphous	Sol–gel, melting	Low strength, low Young's modulus, dental repairs, fillers, middle ear ossicular replacements
Biometals	Ti, Zr, Fe, Mg, Ca, Zn, Sr metals	Biodegradable like Mg and Mg alloys, Fe and Fe alloys and W. Non-biodegradable like Ti and Ti alloys, 316L SS, 304SS, Zr and Zr alloys.	Crystalline	Conventional casting	High strength, high Young's modulus, dental implants, stents, orthopedic surgeries, sutures
Biomedical BMGs	Ti, Zr, Fe, Mg, Ca, Zn, Sr metals	Biodegradable: Mg-based BMGs, Ca-based BMGs, Zn-based BMGs, Sr-based BMGs. Non-biodegradable: Ti-based BMGs, Zr-based BMGs, Fe-based BMGs.	Amorphous	Modified casting	High strength, low Young's modulus dental implants, fillers, stents, orthopedic surgeries, sutures

Table 3.28 BMGs family with composition, preparation, and some mechanical/physical properties [70]

System name	Composition	Preparation methods	Mechanical properties	Corrosion behavior	Biocompatibility
Ti-based	Ranging from Ni and Be: Ti–Zr–Ni–Be, Ti–Ni–Cu–Sn–Be–Zr, Ti–Zr–Cu–Ni–Sn Ni and Be free: Ti–Zr–Cu–Pd, Ti–Zr–Cu–Pd–Sn, Ti–Zr–Si	Arc-melting/copper mold casting, Spark Plasma Sintering (SPS)	Relatively low Young's modulus (80–120 GPa), high fracture strength (1700–2500 MPa), excellent specific strength Shows an elastic elongation of about 2.3% and small plastic elongation of about 0.5%	Exhibits passive behavior at the open-circuit potential with a low corrosion rate, vulnerable to localized corrosion in the form of pitting corrosion Better or equivalent corrosion resistance than 316L stainless steel, CP–Ti, Ti-based biomedical alloys	15 days immersion in SBF[a] is needed to hydroxyapatite (HA) deposits on the surfaces that guarantee the biocompatibility The in vitro cell response (MTT/CCK8 assay and cell morphology observation) and in vivo animal implant of the BMGs demonstrated Better biocompatibility than the conventional crystalline Ti–6Al–4 V and Ti–45Ni alloys
Zr-based	Ranging composition in: Zr–Al–Ti–Cu–Ni, Zr–Cu–Ni–Al–Si, Zr–Cu–Pd–Al–Nb, Zr–Ti–Cu–Fe–Al, Zr–Al–Co, Zr–Cu–Fe–Al–Ag, Zr–Cu–Ni–Al	Arc-melting/copper mold casting, melt spinning, Arc-melting/arc-tilt casting, arc-melting/single roller quenching	Higher yield strength and twice to three times higher in hardness than 316L stainless steel, Ti and Zr-based alloy. In the Ni free form of Zr-based BMGs, yield strength and fracture strength are over 1300 and 1600 MPa, respectively. Plus, the modulus is lower than 316L stainless steel and Ti–6Al-4 V alloy, with about 200 GPa and 110–125GPa, respectively	The passive film mainly composed of ZrO_2 on the surface is to account for the high corrosion resistance. Different physiological investigate like PBS, Hank's solution, 0.9% NaCl saline solution, Ringer's solution, artificial saliva solution, and SBF solutions revealed a lower passive current density, much higher pitting potential compared with 316L stainless steel, Zr and Zr-based alloys and Ti and Ti-based alloys	Better biocompatibility in terms of in vitro (cell culture of MTT/CCK8 assay and cell morphology observation) and in vivo (animal implants) than 316L stainless steel, Zr-based alloys, and Ti-based alloys

(continued)

Table 3.28 (continued)

System name	Composition	Preparation methods	Mechanical properties	Corrosion behavior	Biocompatibility
Fe-based	Ranging composition in: Fe–Co–Cr–Mo–C–B–Y, Fe–Mo–C–B–Er, Fe–Cr–Mo–C–B–Er, Fe–B–Si, Fe–Cr–Mo–P–C–B	Arc-melting/copper mold casting	The Vicker's hardness of FE-based BMGs ranging from 1200 to 1800 MPa, tensile fracture strength of at least 3000 MPa which is more than different alloys. The yield stress could be reaching 2.9GPa with high plastic strains up to 3.6%. By controlling composition, the ductility could be enhanced	Higher pitting potential values, lower corrosion current density values either in Hank's solution or in artificial saliva solution and quiet lower ion releasing than 316L stainless steels	Revealed better biocompatibility for the in vitro murine fibroblast cells (L929 cell and NIH3T3 cell) culture (MTT/CCK8 assay and cell morphology observation)

[a]SBF: Simulated body fluid

Table 3.29 Main properties and preparation methods of Nitinol alloys (Reprinted from International Endodontic Journal, Thompson, An overview of nickel–titanium alloys used in dentistry, no. 302, Copyright (2000), with permission from John Wiley and Sons) [71]

Properties and preparation	55-Nitinol austenite	55-Nitinol martensite
Physical		
Density (gm.cm^{-3})	6.45	–
Melting point (°C)	1310	–
Magnetic permeability	<1.002	–
Coefficient of thermal expansion ($\times 10^6$/°C)	11.0	6.6
Electrical resistivity (Ohm-cm)	100×10^{-6}	80×10^{-6}
Hardness 950 °C (Furnace cooled)	89 R_B	–
Hardness 950 °C (Quenched-R.T. water)	89 R_B	–
Mechanical		
Young's modulus (GPa)	120	50
Yield strength (MPa)	379	138
Ultimate tensile strength (MPa)	690–1380	–
Elongation	13–40%	–
Shape memory		
Transformation temperature (°C)	– 50 to +100	–
Latent heat of transformation	10.4 BTU lb^{-1}	–
Shape memory recoverable strain	6.5–8.5%	–
Super-elastic recoverable strain	up to 8%	–
Transformation fatigue life	several hundred	–
at 6% strain	Cycles	–
at 2% strain	10^5 cycles	–
at 0.5% strain	10^7 cycles	–
Preparation methods		
	Vacuum melting-casting, press forging, rotary swaging and rod/wire rolling	

3.4 Polymers

3.4.1 Introduction

Dental restorative materials have a wide history especially, in terms of polymers. The first polymer application in dentistry dates back to around 1937 by the enhancement of poly (methacrylic acid). Excellent biocompatibility and good mechanical properties led the polymers become famous as dental materials. Generally, polymers could be divided into two classes based on their sources. The first group is natural materials

Table 3.30 Demands for suitable polymers for application as a dental material (Reprinted from Oral Diseases, Lai et al., Advanced functional polymers for regenerative and therapeutic dentistry, no. 555, Copyright (2015), with permission from John Wiley & Sons) [73]

Characteristics and Properties	Remarks
Suitable biomechanical strength	Polymeric materials with weak strength (such as collagen and fibrin) may serve as cell-encapsulating scaffolds, whereas mechanically stiff polymers (such as PLGA and PCL) may be used as structural substitutes
Low production cost	Usage of the polymers requiring high production costs may not be clinically viable and favorable
Low cytotoxicity	Polymers showing comparatively high cytotoxicity may not be applicable clinically
Non-immunogenicity	Polymers used in dental applications should not be immunogenic. This is particularly true for dental implants, or the implant will be rejected by the body
High biocompatibility	Ideal polymers for dental applications should be biocompatible and can undergo biologically safe degradation
Easily processing	Due to the diversity of tooth organs in functions, structures and the spatiotemporal characteristics of tooth bud development, polymers that provide easy processing to form biostructures will be favorable for tooth tissue engineering
Easily storage	Ideal polymers should have a reasonable shelf life after their production, with no harsh requirements being needed to be fulfilled for storage
Acceptable sterilizability	The polymers should be able to be sterilized easily and effectively in the clinical setting
High cell seeding efficiency	The polymers should be able to allow cell encapsulation or provide a surface to which cells can adhere for regeneration of singular or multiple dental tissues

involving chitosan, collagen, agar, and fibrin and the second group is artificial materials which include acrylic resin and its derivatives, polyetherethekeone (PEEK), dendrimer, polylactic acid (PLLA), etc. [72]. As well as other biomaterials, dental polymers should cope with the demands of desirable, minimum needed properties, and Table 3.30 describes these demands.

Polymers have a wide application in dentistry which are in oral medicine (e.g. drug delivery, nucleic acid and therapy, etc.), tooth replacement (e.g. dental filling, dental crown, denture, etc.) and tooth regeneration (e.g. membrane, scaffold, etc.) [73]. Plus, polymeric films (PMFs) can be applied as they are antibacterial/antimicrobial, for corrosion prevention and friction reduction in dentistry. Commonly used polymers for dental application are polyethylene (PE), polymethyl methacrylate (PMMA), polycarbonate (PC), polyethylene glycol (PEG), polydimethylsiloxane (PDMS), polyurethane (PUR), polylactic acid (PLLA), poly(e-caprolactone) (PLC), polypyrrole (PPy), hexamethyldisilazane (HMDC), N-isopropylacrylamide, N-tert-butylacrylamide, Hydrogel, and Polyetheretherketone (PEEK) [74]. However, the most important dental polymers are discussed in the following sections.

3.4.2 *Family of Polymers*

Polyetheretherketone (PEEK). An organic linear, aromatic polycrystalline (can be in two phases of amorphous or crystalline) thermoplastic with white color can be derived from polyurethane (PAEC) substances. This material contains three families based on composition: pure, carbon-containing, or SiO_x glass additions. The structure of PEEK is shown in Fig. 3.30 [75].

PEEK has a soluble temperature of 335–343 °C, maximum operating temperature of 250 °C, and tensile strength of 90–100 MPa. Another capability of PEEK is low friction, resistant to vapors and to sea water, excellent biocompatibility, high chemical resistance in harsh atmosphere, no cytotoxicity, and mutagenicity with human body [76, 77]. PEEK has different applications in dentistry specially in dental implants at which different attempts have been made to improve the cell attachment to the surface of PEEK dental implants. For instance, occupying active gas plasma bombardment for surface modifications with different biological neutral gases like oxygen, nitrogen, and argon [76]. The mechanical properties of PEEK (i.e. tensile strength) are similar to that of bone, dentine, and enamel which leads to less stress shielding consequences. Bio-HPP, one of the modified forms of PEEK has a high anti-allergic, low plaque affinity, and good wear resistance which is a good candidate for RDP framework materials. Table 3.31 shows some incorporation of PEEK with different materials and the outcomes [6].

Polymethyl Methacrylate (PMMA). Has two IUPAC name of poly[1-(methoxy carbonyl)-1-methyl ethylene] and poly(methyl 2-methyl-propenoate) based on hydrocarbon and ester aspects, respectively. It is a transparent thermoplastic, lightweight, shatter-resistance, easily processing conditions, and excellent scratch resistance. The structures of PMMA and MMA are shown in Fig. 3.31 [78].

PMMA is an amorphous polymer with a glass transition temperature (Tg) of 100–130 °C and density of 1.17–1.20 g.cm^{-3} at ambient temperature. Also, the melting point of PMMA is between 105–160 °C with water absorption of 0.3% and tensile strength of 72 MPa. This polymer could be isotactic (mm), yndiotactic, (rr) or atactic (mr) based on the chemical structure. Table 3.32 shows the resistance of food subjection environment [78, 79]

According to Table 3.33, PMMA has wide productive features for dental applications such as satisfying esthetics, easily clinical manipulation, accurate fit, being stable in oral atmosphere, and cost-effective equipment needs. By the way, PMMA currently contains some drawbacks in denture application such as fracture, inflammation caused by the degradation of materials, and adhesion of microorganism.

$$\left(-C_6H_4-OC_6H_4-O-C_6H_4-CO- \right)_n$$

Polyetheretherketone

Fig. 3.30 The structure of PEEK [75]

Table 3.31 Different incorporation of PEEK and the consequences [77]

Incorporation	Results and Comments
Nanometer zirconia	Lowers wear property and friction resistance in comparison with pure PEEK
TiO_2, HAF and HAP NPs[a]	Satisfy the bioactivity of nanocomposites
Carbon fibers	Increases the elastic modulus up to 18 GPa, reduction in the distal stress on the abutment teeth and reduce stress shielding
Surface modification with Tantalum ions	Enhances the elastic modulus to be closer to the cortical bone
PEEK/nano-FHA composite	Better biocompatibility, antibacterial activity, and osseointegration
CFR-PEEK coated by multi-layered nano porous TiO_2	Better elastic recovery, stability, and antibacterial actions
PEEK and PEEK Bioglass particles for Ni–Ti wires	Good adhesion to the metallic substrate with increasing the flexibility

[a]Nanoparticles

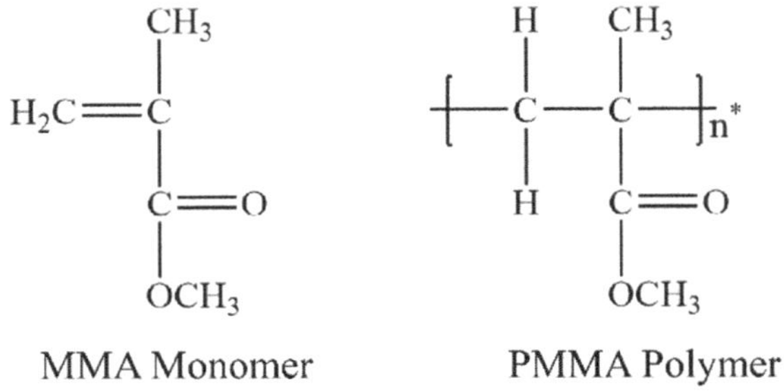

Fig. 3.31 The structure of PMMA and MMA

Table 3.32 Resistance of food subjection environment (Reprinted from Polymer Reviews, Ali et al., A review of the properties and applications of poly (methyl methacrylate)(PMMA), no. 685, Copyright (2015), with permission from Taylor & Francis) [78]

Type of food	20 °C, for one month	50 °C, for one month
Fruit Juice	Not attacked	Not attacked
Milk	Not attacked	Not attacked
Olive Oil	Not attacked	Not attacked
Vinegar	Not attacked	Not attacked
Wine	Not attacked	Not attacked

For solving these problems, three aspects are considerable which are the optimizations of the chemical structures, the admixture with fillers, and alternative materials like polyamide. Table 3.34 indicates different incorporation of PEEK and the consequences [72].

Table 3.33 Different incorporation of PEEK and the consequences [72, 80]

Incorporation	Results and Comments
Reinforcement filler like inorganic oxides and NPs	Enhancement of flexural strength and impact strength
Silver, Zinc oxide and quaternary ammonium salts	Improving antibiotics and dramatic reduction of microorganism's adhesion
S-PRG[a] filler	Prevent dental plaque formation
Surface modification with 16% methacrylic acid and coating with self-bonding polymer	Reduce C. albicans adhesion to the surface

[a]Surface reaction-type pre-reacted glass ionomer

Polyethylene (PE). It is a semi-crystalline thermoplastic polyolefin (POE) which ethylene makes up as single monomer. This polymer could be divided into four main groups. High-density PE (HDPE) contains the most degree of crystallinity and versatile among all the polyethylene family (density ranging from 0.941 to 0.965 $g.cm^{-3}$). The chains of HDPE pack into a crystal structure in excellent formation and lead to be the most stable PE. The second type is low-density PE (LDPE) which is a ductile and flexible material (density ranging from 0.910 to 0.925 $g.cm^{-3}$). The high degree of chain branching attributes for the weak intermolecular interaction and consequently the unique flow properties are provided. The third group of PEs is linear low-density PE (LLDPE) with a large portion of short branches and linear structure (density ranging from 0.919 to 0.925 $g.cm^{-3}$). These properties cause it to be appropriate for film application and provide transparency, high toughness, and flexibility property. The fourth group of PEs is medium-density PE (MDPE) which could be fabricated from different catalyst like Ziegler–Natta, chromium/silica, and metallocene. However, this kind of PE has limited application in medicine (density ranging from 0.926 to 0.940 $g.cm^{-3}$) [81]. Generally, PE's melting point is between 99 °C and 138 °C with a glass transition temperature (T_g) of −20 °C to −128 °C and the tensile strength of 48.6 MPa [79]. The structures of ethylene molecule and PE are presented in Fig. 3.32 [81].

In fiber-reinforced composites (FRC) which provide metal-free adhesive esthetic dentistry, polyethylene fibers application improves the impact of strength, elastic modulus, and flexural strength of FRCs [82, 83]. The polyethylene fibers usage in dltntistry are for perio splints, strengthening removable prosthesis, post and core fabrication, prosthesis, and provisional dentistry [84–88].

Polyethylene Glycol (PEG). On occasion, PEG named as poly (ethylene oxide) (PEO), poly(oxyethylene) (POE), and poly oxirane which generally with the molecular weight of 20,000 belongs to PEG and PEO refers to higher molecular weight. The PEG is viscous, colorless liquid or waxy, white solid at less or higher molecular weights of 1000, respectively [89]. The melting points of the solid (i.e. based on PEG molecular weights, melting point is different such as PEG 400 = 4–8 °C,

Table 3.34 PMFs as antimicrobial agent in different area of dentistry [74]

Antimicrobial agents	PMFs incorporations
Biofilm prevention	Application of polymeric nanomaterial reduces adhesion and biofilm formation. Polymer NPs provide delivery of drugs to the site of biofilm target. Polysaccharides such as alginate, chitosan, and pectin with bioadhesive systems for mouth environment act as electrostatic deposition and bring more stability in saliva environment
Tooth erosion prevention	PMFs stop direct contact of tooth with low pH foods and acidic materials in oral cavity. Polymers such as propylene glycol alginate, highly esterified pectin and gum arabic forming a protective film on the enamel and dentin which drop consequences of acid surface contact. Likewise, chitosan has ability to form a protective layer on enamel and dentin surface near the mucin from saliva. Carbopol is another high-molecular-weight acrylic acid polymer which has a thickening effects application and manages demineralization of enamel
Drug delivery	This process could happen through the mucosa membrane in oral cavity which improves performance of fragile drugs in oral cavity and teeth. Drug delivery and drug formulation processes can be enhanced via polymers and NPs. Chitosan, alginate gelatin, and methacrylic acid are used for polymeric nano-capsules for drug loading. Plus, different pectin (i.e. A high-molecular-weight carbohydrate polymer)-coated liposomes are efficient for their capability to absorb on Hydroxyapatite and dental enamel. Another polymer is polycarbonate micelles for controlling drug-releasing
Restorative dentistry	Polymer adhesive could be applied for dental implant coatings which provide effective biocompatibility and cell adhesive. For instance, dopamine-methacrylate, 2-methoxyetheyl acrylate are used as a dental adhesive because of Fe^{3+}-rich copolymers improved the strength of dental adhesives bonds to an artificial saliva contaminated dentin surface. Other anti-biofilm formation polymers are acrylic acid, alkylmethacrylate, and polydimethylsiloxane copolymer, Carbopol, N-Halamine based polymer additive, and Ti oxide-chitosan/heparin multilayers
Prosthetic dentistry	PMMA with addition of ZrO_2–, Ag- and Pt-NPs reduce bacterial and fungal colonization such as C. albicans biofilm. In addition, phospholipid polymer, poly(2-methacryloyloxyethyl phosphorylcholine-co-n-butyl methacrylate) (PMB) are incorporated into PMMA for reduction of biofilm formation on denture surface
Implantology	Surface modification via polymer/carbon reinforced composites provide acceptable osseointegration and covalent bonds for strong attach to the surface of implant. Other applications of polymer in implant area are the usage of antibiotic-containing nanofiber-based polymeric films via application of bioactive polymers. The most important of the bioactive polymer are hydrogels, hydroxypropyl methylcellulose (HPMC), poly(lactic/glycolic acid) and poly(ε-caprolactone) (PCL) which are also use for drug delivery
Periodontics	2-methacryloyloxyethyl phosphorylcholine (MPC) reduce and control biofilm on tooth surfaces. MPC-ran-2-methacryloyloxyethyl phosphate (PMP) copolymer with zwitterionic and Ca^{2+} is productive for preventing of biofilm formation on tooth surfaces. By the way, the production method of MPC is account for expensive products which is known as a drawback

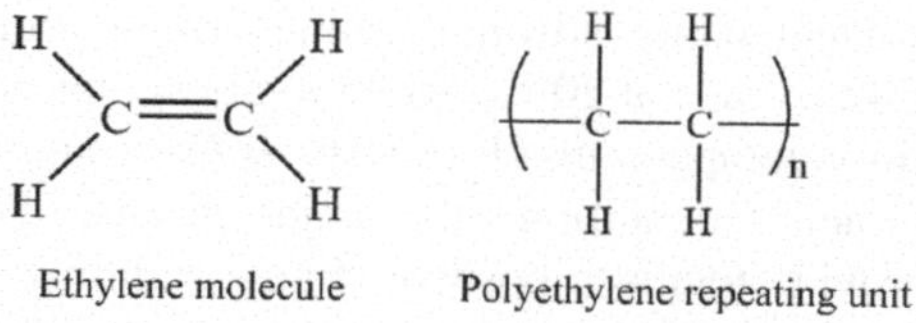

Fig. 3.32 The structure of ethylene molecule and PE

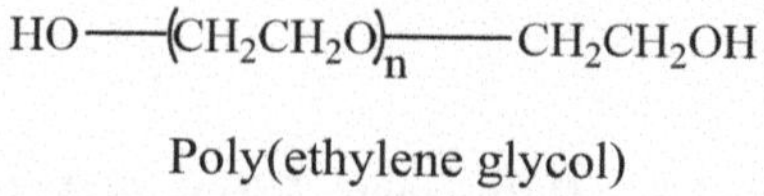

Fig. 3.33 The structure of PEG

PEG 600 = 20–25 °C, PEG 1500 = 44–48 °C, PEG 4000 = 54–58 °C and PEG 6000 = 56–63 °C.) is approaching a plateau at around 67 °C. Additionally, the glass transition temperature (T_g) is between −60 °C and −70 °C [79]. The molecular weights between a few hundred and roughly 20,000 are appropriate for biomedical applications. The structure of PEG is presented in Fig. 3.33 [89].

The density of PEG at 20 °C is between 1.1–1.7 $g.cm^{-3}$ with white color. About mechanical properties, the tensile strength is around 11–60 MPa and for high molecular weight fibers it is between 100 and 200 MPa [79]. This polymer is non-toxic which prevents any harmful impact on proteins or cells besides of the interaction with cell membranes. Additionally, PEG can easily incorporate with other molecules and surfaces in modification process with little consequences. Soluble in different media (e.g. water, toluene, methylene chloride, etc.), insoluble in different environment (e.g. ethyl ether, hexane, etc.), complex formation with metals cations, formation of two-phase systems with aqueous solutions of other polymers and so on that finally make PEG an interesting polymer [89]. One of the advantages of PEG is being an antibiofouling material which provides a hydrophilic environment on a substrate surface. For this reason, the PEG-coating on dental appliance like stainless steel orthodontic wire could reduce adhesion of S.mutans (i.e. mostly leading acidogenic oral bacteria which is account for white spot lesions or demineralization of the tooth enamel) [90]. In other applications, PEG could be applied as a modifier that can decelerate the fast degradation of fibrin because fibrin has poor mechanical properties [72].

Chitosan (CHI). It is the derivative of chitin (i.e. a linear polysaccharide composed of (1–4)-linked 2-acetamido-2-deoxy-β-D-glucopyranose units) which is made by deacetylation and depolymerization of native chitin, partial deacetylation of chitin in the solid state under alkaline conditions like concentrated NaOH. The waste of the food industry especially involving the recovery of carotenoids causes to be a more suitable starting material for chitosan preparation from crustacean shells. Chitosan could be easily prepared in the form of gels, membranes, nanofibers, leads, microparticles, nanoparticles, scaffolds, and sponges [91]. Chitosan is a linear amino polysaccharide with too much nitrogen content, rigid D-glucosamine structure, high hydrophilicity, crystallinity, bio adhesivity and biocompatibility, non-toxic, biodegradable, adsorable, film-forming ability, etc. Chitosan has a density of 1.4–1.42 g.cm^{-3} at 20 °C, white to gray color, melting point of 199–230 °C, glass transition temperature (T_g) of 163–172 °C and tensile strength of 6.7–150.2 MPa [79]. In acidic conditions, the amino groups of chitosan capture hydrogen ions which provide totally a positive charge resulting in bioadhesive capability to negatively charged surfaces such as tooth enamel, soft tissue, and so on [92]. The structures of chitin and chitosan are shown in Fig. 3.34 [91].

3.4.3 Polymeric Films (PMFs) in Dentistry

Besides metal and ceramic dental materials, polymeric films (PMFs) are being most common in dentistry because they do improve not only mechanical and physical properties but also enhance antibacterial actions in the oral cavity environment as restorative, preventative, or regenerative therapies. PMFs serve different applications which according to Fig. 3.35 can be divided into an agent for reduction in corrosion, reduction in friction, and increasing antimicrobial actions [74].

Due to the dental environment and acidic foods, different applied restorative materials (e.g. Ni-Ti alloys, stainless steels, etc.) are susceptible to corrosion attack. For this, biocompatible PMFs are more attractive as coating on restorative materials. These PMFs include polypyrrole (PPy)/HA nanocomposites, polyurethane (PUR), polyamide, PEEK, polytetrafluoroethylene, and hexamethyldisilazane (HMDS). In addition, PMFs could cover the defects from processing which decreases the dental decay propagation. The drawbacks of PMFs, however, are the roughness or detach

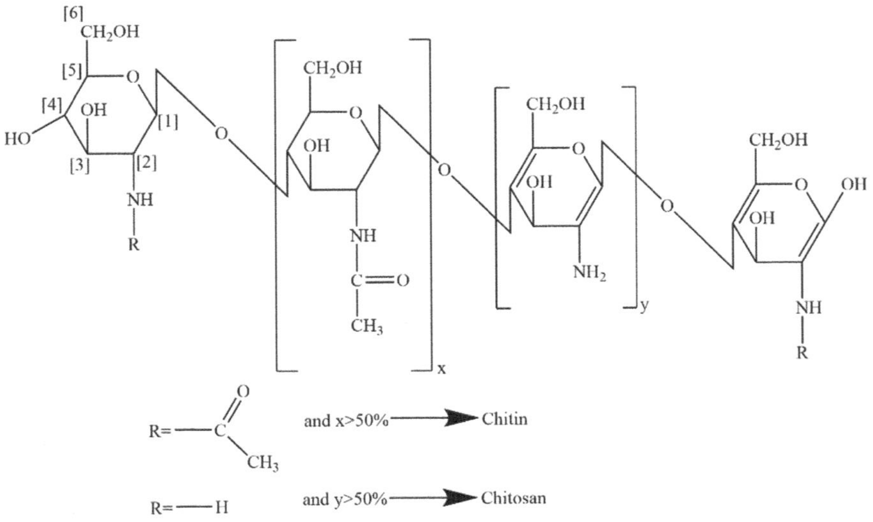

Fig. 3.34 The structures of chitin and chitosan [91]

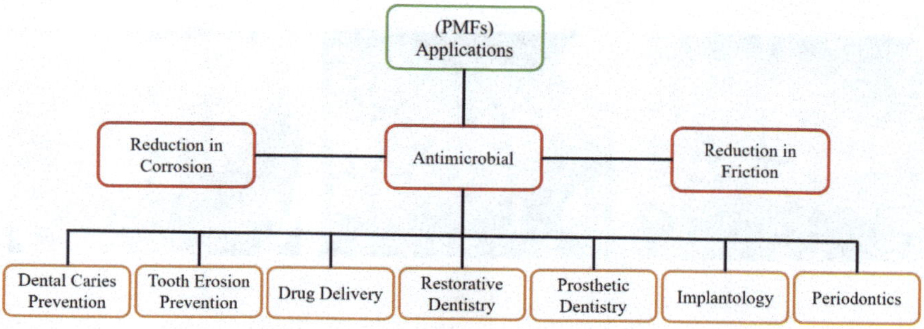

Fig. 3.35 The application of polymeric films (PMFs) in dentistry [74]

creation after long-term usage on the surface like Ni-Ti wire. Likewise, the application of conductive biocompatible 3, 4-ethylenedioxythiphene films dramatically declines the corrosion attacks of Mg-based medical implants. Polyamide (PA) could seal Ni-Ti wire which drops the nickel ion releasing of around 85% and corrosion actions [74].

Mastication and eating foods drive different forces on restorative materials and implants which should be considered in material designing. For instance, application of polyamide (PA) coating on Ni–Ti wires could decrease the wear rate and dynamic friction coefficients in comparison with uncoated wires. Another polymer is in the form of hydrogels that are used in dentistry like coatings on Ni–Ti implants and orthodontics due to the similar mechanical properties to many natural tissues and their viscoelasticity. Different mixture of polymers could provide attractive properties like reduction in friction. For instance, incorporation of poly-vinyl alcohol with different amount of zwitterionic polymer film, poly ([74-(methacryloyloxy) ethyl] dimethyl-(3-sulfopropyl) ammonium hydroxide) show friction reduction of zwitterionic polymers of around 80% which is suitable for dental material application.

PMFs have application in dentistry as an antimicrobial agent which prevents the common problem of microbes adhering to teeth. Besides other techniques like adding antibacterial nanoparticles (NPs) and so on, antimicrobial polymer films (e.g. acrylic acid, alkylmethacrylate and polydimethylsiloxane copolymers) are applicable for that demand. Table 3.34 shows other applications of PMFs based on antimicrobial agent [74].

3.4.4 Shape Memory Polymers (SMPs)

Shape memory materials like SMPs are stimuli-sensitive which can respond to the environment changes such as temperature, light, magnetic field, or mechanical strength application. For instance, thermal-responsive polymers or thermally induced shaped-memory polymers are changing the shape by fluctuation of temperature. Shape memory properties are based on different factors like polymer structure

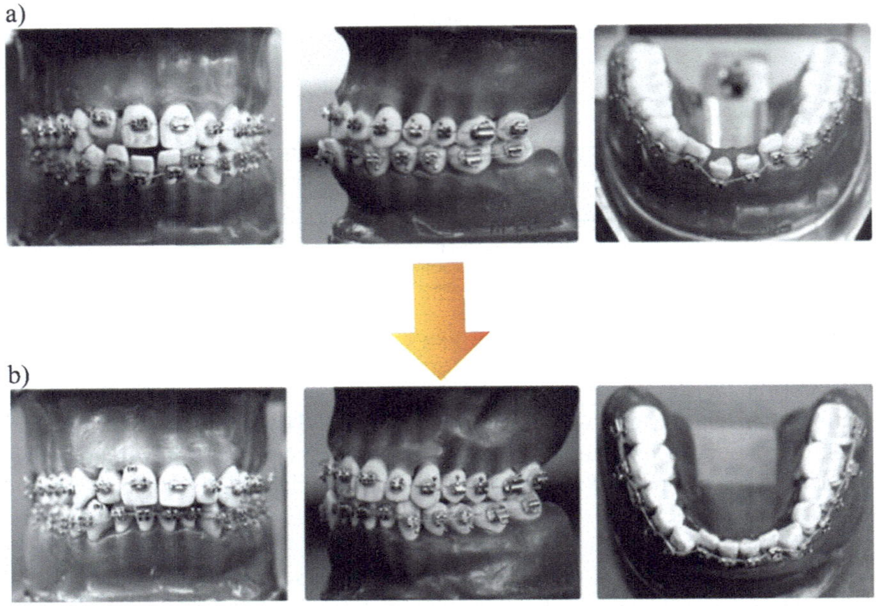

Fig. 3.36 Photograph of orthodontic wire produced by PUSMP. **a** before and **b** after orthodontic treatments. (Reprinted from In Shape Memory Polymers for Biomedical Applications, do Nascimento et al., Shape-memory polymers for dental applications, no. 274, Copyright (2015), with permission from Elsevier) [94]

and morphology in parallel with processing and programming technology. For understanding the SMPs behavior, thermally induced SMPs are described. The polymer firstly is conventionally processed to catch its permanent shape. Then, the SMPs are deformed and received a temporary shape, which is called the programming (i.e. include heating up the specimen, deformation, and cooling the specimen or drawing the specimen at a low temperature). The permanent formation is programmed when showing the temporary shape. $T_{trans.}$ is the temperature of transition and any temperature more than that causes the shape memory behavior. Finally, the recovery of the programmed SMPs happened. Cooling the temperature below the $T_{trans.}$ causes the solidification of the material at which no recovery of the programed SPMs is observed [93]. SMPs are used in dentistry for anticaries and antibacterial activity. They could be prepared in different ways like one-step polymerization of monomers/prepolymers and cross-linking agents, chemical cross-linking of high molecular weight thermoplastic polymer, mixing of various polymers directly, and one-step synthesis of phase-segregated block copolymers. Thermally induced SMPs expand and contract by temperature altering which decrease the gap formation at the interface, consequently, decrease microleakage. For instance, polyurethane block copolymer (PU) is used to be prepared as shaped-memory wire for orthodontic usage. This SMP arranges the misaligned teeth by body temperature altering. Moreover, it was made with soft and hard parts, which bring enhanced flexibility and mechanical resistance of orthodontics wires. Figure 3.36 shows a photograph of an orthodontic wire made by PUSMP [94].

3.5 Composites

3.5.1 Introduction

Composites are solid materials containing at least two distinct materials or phases. Properties of composites are different from those of each constituent material or phase and are generally an average of the properties of the constituents. Enamel and dentine are examples of natural composites. Glass ionomers (GIs), resin-modified glass ionomers (RMGIs), and polymer matrix composites (resin composites) are the composite materials used in direct restorative dentistry. Due to the water-based nature and acid–base setting reaction of GIs and RMGIs, these materials are generally individually classified as cement restorative materials [3]. Dental composites (resin-based composites) consist of an extremely cross-linked monomer resin (matrix) and inorganic or polymeric particles mainly mentioned as fillers bonded to the matrix via a coupling agent, while the composite resin contains only polymeric fillers. Up to the past decades, amalgam was the most convenient restorative material, however, its inappropriate color, and mercury content has replaced its application with more esthetic ceramics and resin-based composites [1]. Mechanical properties of dental composites such as flexural strength, fracture toughness, and tensile strength are generally comparable to those of amalgam and porcelain restoratives, however, their elastic modulus is much lower than amalgam [95]. Resin-based composites by the side of compomers, metals, metal-ceramics, ceramics, and polymers are among the materials extensively used for restoring or replacing injured tooth structures in dentistry. Resin-based composites are used for various dentistry applications such as cavity liners, cores, buildups, inlays, onlays, and provisional restorations [95], as well as sealants of pits and fissures. Composites are also used for cementing crowns, bridges, fixed dental prostheses, and ceramic veneers [1]. Composites are the best choice for restoring anterior class V caries and are currently widely used for restoring posterior teeth in many countries [95]. Utilization of composites as filling materials has largely improved restorative dentistry owing to various outstanding properties of these materials such as good esthetics, easy manipulation, and handling properties (due to their high range of fluidity or rigidity), and wear resistance to be used for replacing injured teeth [1, 96]. Through the past five decades, several developments have been performed in restorative materials exhibiting appropriate esthetics. Silicates were the first restorative materials with good esthetics used for filling cavities or preventing their progress, which have been limited in use due to their high rates of erosion in the oral environment. Afterward, acrylic resins with a color similar to natural teeth replaced silicates owing to their low rates of solubility in oral environment, low cost, and easy manipulation. These materials present major drawbacks related to their inadequate wear resistance, high shrinkages through curing process, and great thermal expansion and contraction coefficients increasing the risk of leakage. To reduce these disadvantages, quartz particles, which do not shrink and have a low thermal expansion coefficient at ambient temperature, were added to the composition forming composite materials. The most crucial factor, which restricted the use of these primary composites, was the lack of proper bonding between fillers and resin matrix. Bowen in 1962 was among the first who introduced a durable composite made up of bisphenol-A glycidyl dimethacrylate (*bis*-GMA) and an organic silane as a coupling agent to overcome disadvantages of the previously introduced dental composites. Since 1962, many developments have occurred regarding the fillers by reducing their size to enhance esthetics and polishability of dental composites. Besides this, barium glass and mineral fillers have been applied to improve radiopacity, manipulation, curing shrinkage, and mechanical properties of these composites. Macrofilled composites, which are nowadays called traditional composites, were introduced in the 1970 s. Macrofilled composites contain large and ground particles of quartz or amorphous silica. Ground fillers enhance mechanical properties, reduce water adsorption, and decrease curing shrinkage as well as thermal expansion of the composite. One still existing disadvantage of these kinds of composites is their great roughness due to the high rates of resin matrix abrasion. Aiming to improve the surface conditions of these composites, smaller inorganic filler particles with wide size distributions ranging between 0.5 μm and 3 μm were used. The resulted composite, which was made up of 80–90 wt% fillers presented smooth surface, high wear resistance, and decreased curing shrinkage. Between the decades 1980 s and 1990 s, much attention has been paid to the filler development resulting in the introduction of microfilled, and hybrid composites. Light-cured composites were introduced in the 1980 s as a result of the progress in curing technology with improved working time and manipulation. In this manner, ultraviolet (UV) curing systems were the first family of light curing systems, which faced several drawbacks resulting in its replacement by visible blue light curing systems [1]. From the early 1990 s to the 2000 s studies were carried out with the aim of decreasing polymerization shrinkage and that enhanced clinical handling of composites leading to introduction of flowable and packable composites. In the past decade, bulk-fill composites have received much attention in order to reduce clinical time required for restoring teeth with composites as well as enhancing antibacterial and longevity of composites. Self-adhesive, fiber-reinforced, antibacterial, self-healing, and remineralizing composites have been studied and developed in this decade (Fig. 3.37) [97].

Introduction of some novel monomers such as UEDMA and BisEMA besides novel initiation systems and fillers have developed the desired properties of composites and have extended their application as both direct and direct restorations [98]. At the present stage of advanced dentistry, application of resin-based composites has increased greatly conducing total replacement of mercury-containing amalgams for direct restorations by them in some countries. The widespread application of

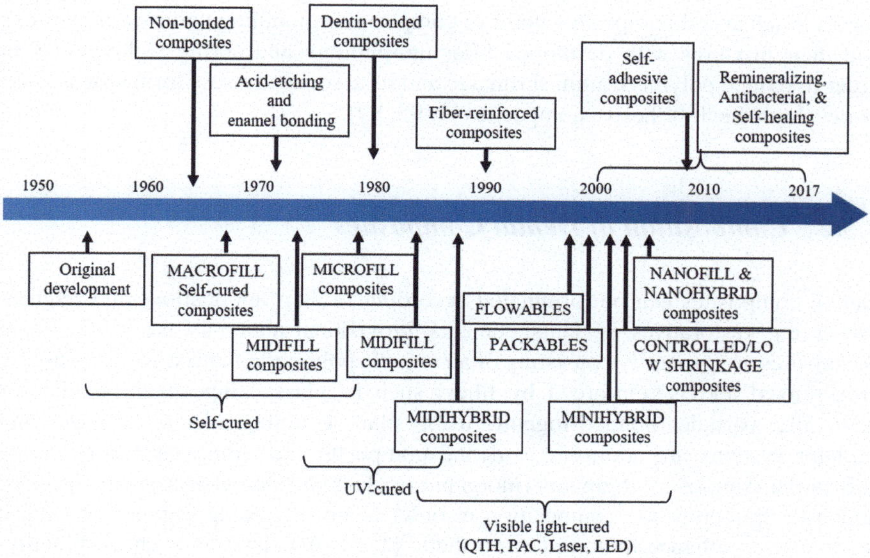

Fig. 3.37 Timeline of the history and development of dental composites [1, 97]

resin composites is owing to their outstanding esthetics, great mechanical properties, and chemical match with natural teeth as well as their capability to protect and conserve healthy natural structures due to their retention method, which is more dependent on luting methods than cavity forms [99, 100]. Nowadays, various types of resin composites are commercially available for dentistry applications such as non-shrinkage composites, flowable composites, and packable composites, which have solved the problems of stress-related debonding, high internal porosity, and high water absorption of traditional composites [101]. A composite application is largely technique-sensitive and requires a careful consideration of the composition, procedure, and bonding cements [100]. The most crucial clinical drawbacks of composites are polymerization shrinkage leading stress transmission to remained natural enamel and dentin, tooth sensitivity after operation, marginal staining, and secondary caries. On the other hand, the biocompatibility of dimethacrylate monomer and initiators is under discussion [99]. Composites are very promising alternatives of restorative materials. Until now, many efforts have been done to enhance composite properties by incorporating calcium fluoride-containing fillers, polymer nanofibers, fused silica whiskers, glass fibers, titania nanoparticles, and dicalcium or tetracalcium phosphate nanoparticles. Incorporation of antibacterial and remineralizing agents such as fluoride, chlorhexidine, zinc oxide nanoparticles, quaternary ammonium polyethyleneimine nanoparticles, and MDPB monomers into dental composites have also been studied [95]. Addition of nanofillers has provided composites with more wear resistance and better esthetics. Besides these, the development of new curing technologies has made composites fortunate restorative materials [101]. The latest trends in advanced composites point to composites containing complex monomers and their mixtures with developed filler morphology and curing technologies in order to reduce polymerization shrinkage and its associated stress formation as well as developing self-adhesive composites [1, 95, 99].

3.5.2 Composition of Dental Composites

Dental composites can be formulated according to their applications and requirements (e.g. restoratives, sealants, cements, provisional materials, etc.) [95]. Resin composites are generally consisting of an organic polymeric matrix, which is highly cross-linked and is reinforced by fillers such as glass, silica, metal oxides, or resin filler particles bonded together using silane coupling agents. Activator and initiator systems are additives of dental composites providing hardening of the soft initial composite. Pigments (inorganic oxides) and fluorescent agents are also added to the composite composition in order to mimic the appearance of natural teeth, and to enhance its long-term esthetic [1, 3]. Accelerators such as dimethylaminoethyl methacrylate (DMAEM), ethyl-4-dimethylaminobenzoate (EDMAB), and N,N-cyanoethyl-methylaniline (CEMA) are the compounds acting on initiator system to proceed curing process through appropriate clinical duration. Stabilizer or inhibitor systems (e.g. hydroquinone monomethyl ether) are also added to dental composite formulation in order to minimize spontaneous polymerization and prolong storage life of products before curing followed by their stability after curing due to their high reactivity with free radicals [1, 100]. Additionally, the presence of ultraviolet absorbers such as 2-hydroxy-4-methoxybenzophenone improves the color stability of the product through its lifetime [100].

3.5.2.1 Polymeric Matrix

The resin matrix of dental composites is commonly composed of aromatic and aliphatic dimethacrylate monomers such as Bis-GMA and urethane dimethacrylate (UDMA) with large molecular volumes providing low polymerization shrinkage (about 0.9%) [1]. Both of these monomers contain carbon double bonds at the end of molecular chains. Free-radical initiation causes the double bonds of these monomers to go through addition polymerization. The refractive index of these aromatic monomers corresponds to that of radiopaque glass fillers resulting in the good optical properties of these composites [3]. As these monomers are strongly viscous, it is essential to blend them with other monomers having lower molecular weights in order to facilitate blending and manipulation of the composite. Triethylene glycol dimethacrylate (TEGDMA), bisphenol-A ethoxylate dimethacrylate (Bis-EMA6), aliphatic urethane dimethacrylates (UEDMA), partially aromatic urethane dimethacrylate, and highly branched methacrylates are examples of monomers with lower molecular weights [3, 102]. A major drawback of the use of these diluent monomers is their high polymerization shrinkage, which increases the risk of leakage in marginal gaps [1]. Low-shrinkage composites containing methacrylate and epoxy (silorane) monomers have been recently introduced in order to decrease the volumetric shrinkage and internal stress of composites formed during polymerization. Silorane monomers contain hydrophobic siloxane groups and ring-opened cross-links of oxirane groups. In low-shrinkage methacrylate monomers, the larger distance between the methacrylate groups lowers the density of cross-links [3]. Figure 3.38 illustrates the chemical structure of monomers utilized in the composition of resin composites.

3.5.2.2 Fillers

Fillers are imported in the structure of the resin matrix in order to enhance the physical and mechanical properties of the organic phase by reinforcing the resin matrix, providing radio-opacity, improving handling properties, and decreasing curing shrinkage, and thermal expansion coefficient [1, 100]. In general, the use of fillers in the composition of dental composites provides six major benefits, which are (I) reinforcement and increase of clinical performance, (II) reduction of polymerization shrinkage (from about 1 vol.% to 4 vol.%), (III) reduction of thermal expansion and contraction coefficients approaching to that of natural teeth, (IV) workability and viscosity control, (V) decrease of water adsorption, which lowers the risk of abrasive wear and staining, and (VI) imparting radiopacity to the radiolucent resins in order to diagnose various kinds of teeth injuries achieved by the addition of heavy metal-containing glass fillers [1]. Composite materials may contain one or more inorganic fillers dispersed within the matrix with large portions [3]. Finely ground glass or quartz, aluminum silicate, lithium aluminum silicate, ytterbium fluoride, and barium, strontium, zirconium, and zinc glasses are the examples of filler particles used in dental composites. Quartz is a filler material, which has been used for a

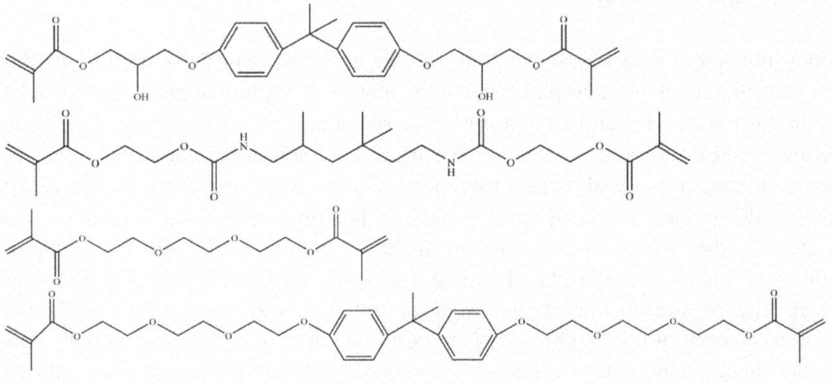

Fig. 3.38 Chemical structures of (1) Bis-GMA, (2) UDMA, (3) TEGDMA, and (4) Bis-EMA6 [3]

long time presenting chemical stability and high hardness, which is hardly grinded or polished and can cause abrasion of the opposing teeth. In contrary, amorphous silica with the same composition and refractive index as quartz has lower hardness and causes reduced amount of opposing-teeth abrasion. Glass and quartz fillers have refractive indices almost close together (about 1.5), which match to that of the resin matrix (about 1.55 for bis-GMA and 1.46 for TEGDMA), hence, providing appropriate translucency similar to that of natural teeth [1]. Recently, nanoparticles are also added to the resin matrix [101]. Higher filler contents generally bring about stiffer composites with higher elastic modulus, higher fracture toughness, better translucency, radiopacity, polishability, and higher wear resistance, which are totally desirable [101, 102]. In order to have a composite with higher values of filler loading, it is essential to have fillers with a broad particle size distribution, which is relied on the use of fillers with smaller sizes. Besides this, utilization of small particles provides smoothness and esthetics to dental composites [1]. The particle size of fillers incorporated in the dental composites has been decreased through the years [102]. Today, filler particles with sizes ranging between 0.005 and 2 μm are employed. Employing particle sizes greater than the visible light wavelength causes opacity and roughness of the composite, which are not desirable. On the other hand, small particles result in viscosity increase due to their superior surface-to-volume ratio, which increases the amount of polar and hydrogen bonds. In this case, the filler content should not excess the amount of about 80 vol.%. As a conclusion, the balance between esthetics, ease of manipulation, and durability should be considered for selecting the used components [1]. By the development of nanotechnology, novel resin composites with enhanced properties containing high loadings (up to 79.5%) of zirconium/silica or silica nanoparticles with particle sizes below 100 nm have been produced. Utilization of nanoparticles decreases curing shrinkage, reduces deflection of cusp walls, and minimizes the possibility of microfissures formation, which may lead to marginal leakage, staining, bacterial penetration, and postoperative sensitivity [100].

3.5.2.3 Coupling Agents

The coupling agent is a bonding agent, which is applied on the surface of filler particles to enhance their chemical bond to the matrix. Coupling agents play a crucial role in the formation of chemical bonds between filler particles and resin matrix. They are difunctional compounds with active surfaces, which adhere to filler particles besides their coreaction with resin monomers. These compounds provide increased physical and mechanical properties as well as leaching inhibition. Titanates and zirconates are the least utilized coupling agents, whereas γ-methacryloxy-propyl-trimethoxy silane (A174 silane) is the most important one used in dental composites [1]. Silane has been used in composites for seventy years ago and is only utilizable with materials containing Si-O bonds (for example silica in composites) [101]. The functional mechanism of this coupling agent consists of hydrolyze of the methoxy groups (–OCH3) to silanol (–Si–OH) groups in the vicinity of water. Silanol groups of the coupling agent and those present on the surface of silica are able to bond

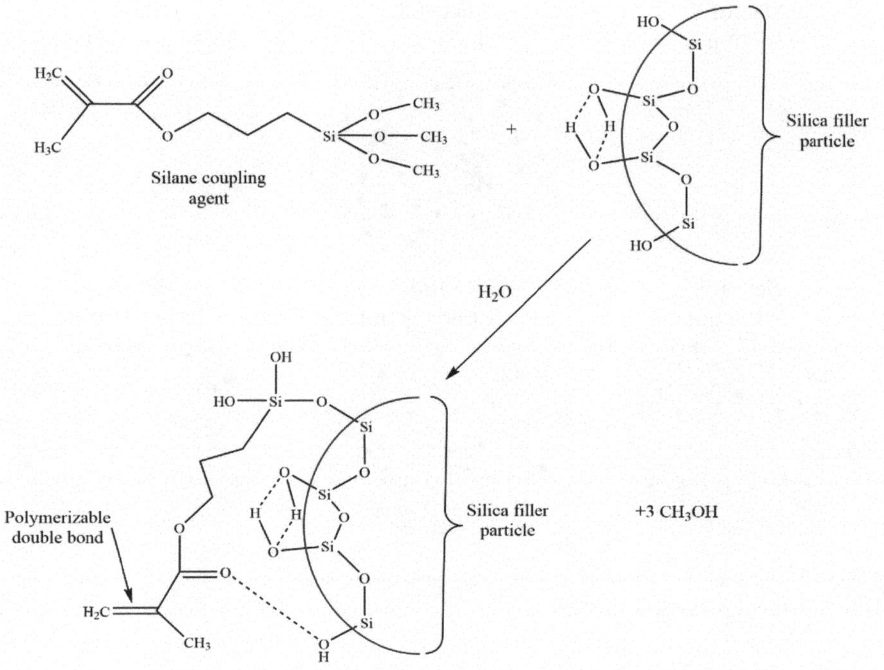

Fig. 3.39 An illustration of hydrolyse of γ-methacryloxypropyltrimethoxy silane making chemical bonds with silica filler particles [1]

together chemically through ether bridges (formation of siloxane bonds, –Si–O–Si–). The coupling process is completed by the formation of covalent bonds between the organosilane methacrylate and polymerized resin (Fig. 3.39) [1, 101]. Formation of chemical bonds between resin matrix and filler particles leads to stress transfer, which deviates cracks to the strong fillers. Despite outstanding wettability of silane, this compound is prone to dimerize or trimerize, which prevents its function as coupling agent. The exact chemical interactions around filler particles and the function of coupling agents is a topic that has been barely studied [101].

3.5.2.4 Initiator-Accelerator System

The initiator-accelerator system is used to polymerize and cross-link the polymeric matrix in order to obtain a hardened structure. This process can be initiated chemically (within self-curing composites), photochemically (within light-activated composites), or by both of them (within dual curing composites) [3].

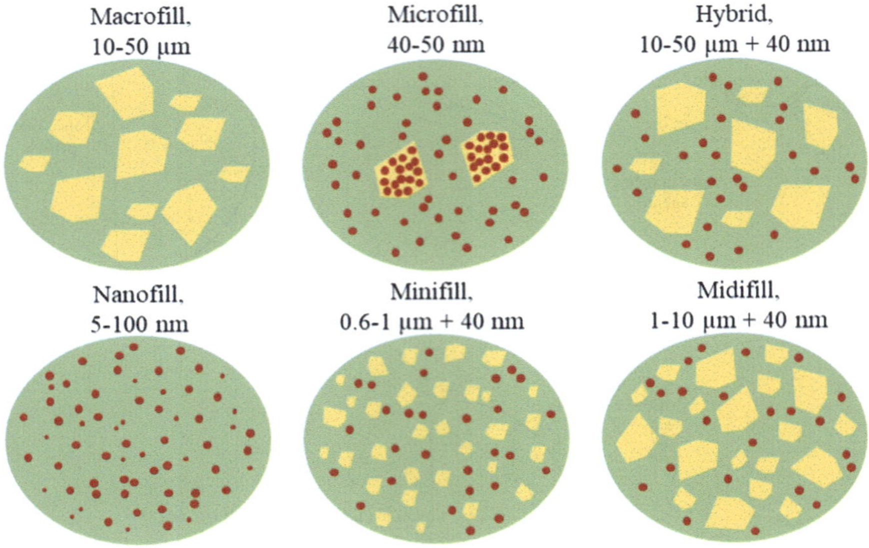

Fig. 3.40 Illustration of various types of dental composites on the basis of their filler particle size. Redrawn from Ferracane et al. [95]

3.5.3 Classification of Dental Composites

Dental composites can be classified in terms of several characteristics such as elastic modulus, inorganic filler content, surface roughness, and compressive strength [100]. There are two mostly used classifications of dental composites based on the particle size of reinforcing fillers and manipulation characteristics of the monomer matrix. The most practical classification of dental composites is on the basis of the particle size of fillers (Fig. 3.40). Using this classification, hybrid composites encounter overlaps due to their special nature that is composed of two or more particle sizes. Nowadays, advanced dental composites are composed of filler particles with sizes less than 0.5 μm accompanied with more than 10 wt% micro or nanofillers providing appropriate viscosity and clinical manipulation [1].

3.5.3.1 Macrofilled Composites

Conventional dental composites, also called macrofilled composites, are the earliest type of composites containing large particles (20–30 μm) with spherical or irregular shapes. These composites are strong, almost opaque, and present low wear resistance [3, 95]. Due to the presence of large filler particles within macrofilled composites achieving smooth surfaces is rather impossible [95].

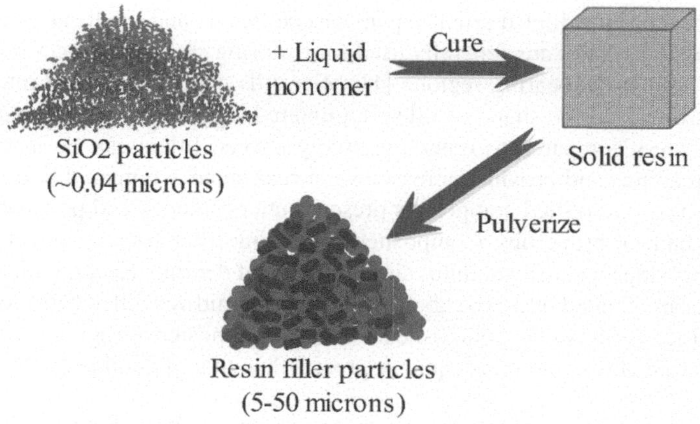

Fig. 3.41 Preparation process of resin filler particles used in microfilled composites. Redrawn from Anusavice et al. [1]

3.5.3.2 Microfilled Composites

Microfilled composites are composed of inorganic colloidal silica agglomerates and resin fillers inserted in a polymeric matrix with particle sizes ranging between 0.01 μm and 0.1 μm and 5 μm to 50 μm, respectively [1]. The name of these composites was selected inappropriately at that time due to the presence of nanoparticles within these composites. In fact, microfilled composites are nanocomposites containing amorphous silica particles with particle sizes about 40 nm. These composites present better polishability and esthetics than macrofilled composites; however, due to the low filler contents their strength is rather poor [95]. The small particle sizes of colloidal silica provide translucency and surface smoothness to these class of composites. By the way, due to the high surface-to-volume ratio of the filler particles and their superior tendency to form agglomerates, the clinical manipulation of these composites is rather complicated. In this case, high concentrations of resin (60–70 wt%) in the composition lowers the viscosity of the composite. The silane-treated silica-containing resin is then cured and pulverized to obtain resin powder with particles ranging between 5 μm and 50 μm. The obtained powder is then used as an organic filler of the heterogeneous microfill composite. In this case, the resin particles are composites containing pyrogenic colloidal silica particles with diameters between 0.01 and 0.1 μm. Figure 3.41 illustrates the preparation process of fillers of the microfilled resins. Although microfilled composites contain small amounts of inorganic fillers, they present low curing shrinkage due to the low shrinkage of composite particles. One of the important shortages of these composites is the weak interface between the microfill particles and cured resin matrix, which reduces their wear resistance. As a matter of fact, microfilled composites are not a good choice for restoring stress-bearing regions. Due to the high contents of resin (40–80 vol.%) presented in the composition of microfilled composites, they exhibit high water adsorption, high thermal expansion coefficient, and small elastic modulus. Microfilled composites are generally used for restoring smooth surfaces (classes III and V) in low stress-bearing regions [1]. Microfills can be classified into several subclasses based on the shape of fillers (splintered, agglomerated, spherical, etc.). Spherical fillers compared to other irregularly shaped fillers can be incorporated in higher loadings and provide increased wear resistance. Compared to other types of composites, microfilled composites present better esthetics and polishability but lower mechanical properties. Composites containing filler contents equal to about 60 vol.% provide an elastic modulus similar to that of dentine. Keeping this in mind, composites are divided into two subclasses including midway-filled (<60 vol.%) and compact-filled (>60 vol.%) composites. For restoring posterior regions, composites should have an elastic modulus equal or greater than that of dentine [102].

3.5.3.3 Hybrid Composites

Particle size decrease of the conventional composites resulted in the production of "small particle hybrids", which were also called minifilled composites composed of microfine particles (0.01–0.1 μm) accompanied with fine particles (0.1–10 μm) in order to achieve smooth surfaces and appropriate mechanical properties [1]. Followed by further developments, composites with submicron particles (0.4–1.0 μm) called minifilled composites (microhybrids) were introduced with enhanced strength and polishability to be used in both anterior and posterior restorations [95]. Microhybrid composites contain fine particles (finely grinded glass, quartz, or ceramic) with particle size ranging between 0.04 μm and 1 μm mixed with microfine silica. These composites contain 77 wt% to 84 wt% filler content achieved due to the wide filler particle distribution. The major drawback of hybrid and microhybrid composites is the loss of surface smoothness during time [3]. Addition of nanoparticles to the composition of microhybrids has led to the production of nanohybrids by several producers. Actually, it is difficult to separate microhybrid and nanohybrid composites since their composition, flexural strength, and elastic modulus are almost similar [95]. Hybrid composites present various benefits such as wide color and translucency ranges, low curing shrinkage, low water absorption, high polishability, similar-to-tooth wear, abrasion, and thermal expansion coefficient [100]. Hybrid composites are employed for restoring high stress-bearing regions, which also require good esthetics such as incisal edges and small non-contact occlusal cavities of anterior regions like class IV regions [1].

3.5.3.4 Nanofilled Composites

Utilization of nanotechnology in dental composites is the most recent advance in the field of restorative dentistry [95]. Nanofilled composites (nanocomposites or nanohybrid composites) are a kind of composites containing nanoparticles (1–100 nm). Oxide nanoparticles are the commonly used materials in dental composites. Due to the nanometric sizes of filler particles, nanofilled composites present higher polishability and optical properties than other types of composites. Non-agglomerated monodispersed nanoparticles and nanoclusters are two types of nanofillers used in nanocomposites (Fig. 3.42). Nanoclusters can be fabricated in wide size distributions providing better rheological properties to nanocomposites compared to that of monosized nanoparticles [3]. The surface of nanoparticles is usually treated using coupling agents such as silane or γ-methacryloxypropyltrimethoxysilane followed by their embodiment into macromolecule chains, which inhibits agglomeration of nanoparticles. Nanoclusters form three-dimensional agglomerates with increased viscosity while the nanometric fillers are discrete presenting composites with lower viscosities. Filler loadings more than 78 wt% provide desirable mechanical properties of restorations applied in high stress-bearing posterior regions. As some of the nanoparticles form loose agglomerates with sizes reaching to even 1.4 microns and due to the lack of chemical bond between these agglomerates, these agglomerated composites present lower mechanical properties than hybrid or microfilled composites. Aiming to overcome this problem, fine glass or pre-cured nanoparticle-filled resin organic filler particles with larger particle sizes are added to the nanofilled composites. As a matter of fact, most of the nanocomposites are actually hybrid nanocomposites and the products containing isolated, discrete, and homogenous nanoparticles have not yet produced commercially [1]. 3 M utilizes the sol–gel method to produce nanospheres, which can form nanoclusters by agglomeration. 3 M ESPE Filtek Supreme combines submicron particles with nanoclusters to fabricate hybrid composites [101]. Besides, shape, amount, and size of filler particles play a major role in the properties of composites. However, since nanohybrid composites are composed of conventional particles accompanied with nanofillers, their performance and category are rather unknown [103]. Several studies have been performed on the clinical performance of nanofilled and nanohybrid composites with comparison to microhybrid composites in the posterior regions and few studies have assessed the comparison between these composites applied in the cervical and anterior regions.

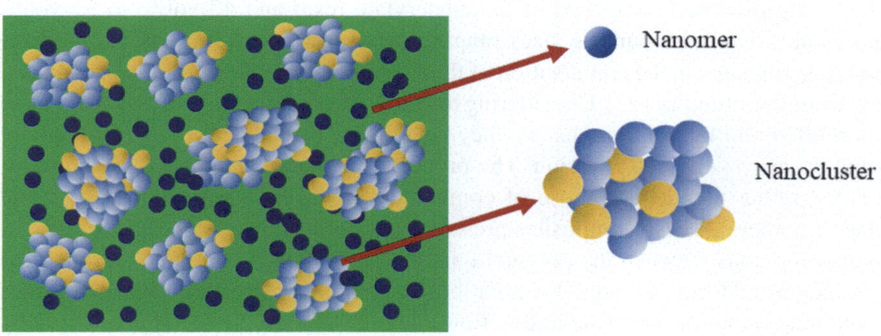

Fig. 3.42 A schematic of a composite containing both nanometric fillers and nanoclusters. Redrawn from Powers et al. [3]

Several studies have shown that in the posterior regions, both nanofilled and nanohybrid composites provide appropriate clinical performance similar to microhybrid composites. In fact, nanofilled and nanohybrid composites do not present any significant benefit over microhybrids regarding surface condition, marginal quality and wear or fatigue resistance [104]. Moraes et al. have assessed the clinical performance (tensile strength, surface roughness, water adsorption, and solubility) of conventional nanohybrid resin composites in comparison with nanofilled and microhybrid composites in order to classify nanohybrids within either nanofilled or microhybrid classes. Their results have shown that nanohybrid composites mainly present unsatisfactory properties compared to the nanofilled composites. On the other hand, their properties are similar or slightly better than microhybrid composites. They have concluded that the properties of nanohybrid composites are much more similar to microhybrid composites than nanofilled composites and they are not advanced novel materials, mostly popular because of the marketing plans [16]. On the other hand, in spite of the fact that nanocomposites present superior flexural strength compared to microhybrid composites, their storage life is not so convenient [102]. After all, studies conclude that nanofilled and nanohybrid composites appear to be convenient choices to be used instead of traditional microhybrid composites and it is up to dentists to choose between the proposed restorative materials for each case [104].

3.5.3.5 Flowable Composites

Flowable composites, also called syringeable composites, have gained attention since 1995 and were produced after progress done over small-particle and hybrid composites. Flowable composites usually exhibit low viscosity and good flowability, which facilitates production of restorations, bases, and liners with desired shapes and easy clinical manipulation particularly in posterior regions where the accessibility is limited [1]. Flowable composites flow easily through fine bores of syringes and provide better adaption in the restoration of small cavities [95]. Flowable composites are composed of dimethacrylate resin and 42 vol.% to 53 vol.% inorganic fillers with particle sizes ranging between 0.4 and 3.0 μm [1]. Utilization of surfactants in the composition of these composites reduces their viscosity and increases their fluidity [95]. Considering that flowable composites are less sticky than microfilled and hybrid composites, they have lower mechanical properties and are more prone to wear and abrasion. The properties and application of these composites are rather identical to those of compomers (hybrids of resin composites and glass ionomers). These composites present pseudo-plasticity due to the presence of hydrogen bonds between the polymeric matrix and fillers, which break through the syringe tip and form again quickly after its placement. Recently, flowable composites containing nanofillers (self-adhesive flowables) have been produced commercially [1, 3]. Flowable composites are prospective restorative materials due to their high wettability of the tooth, excellent penetration, capability of forming of thin layers, significant flexibility, high radio-opacity, and availability in different color shades. These composites are applied in class V restorations, class I and II cavity liners, minimal occlusal restorations, and fissure sealants due to their great adaptability and flowability. Nonetheless, the major drawbacks of flowable composites are high curing shrinkage, and poor mechanical properties [1, 100]. Flowable composites with adhesive monomers are a new branch of dental composites containing acidic monomers including glycerolphosphate dimethacrylate (GPDM). At this time, these newly introduced composites are applied for liners and small restorations, and are looking forward to find their place in many other dentistry applications [95].

3.5.3.6 Condensable (Packable) Composites

Employment of composites is much more difficult and time-consuming than amalgam due to their high plasticity, which inhibits their vertical packability. Packable composites contain elongated filler fibers with sizes of about 100 μm providing desirable strength and handling characteristics. Utilization of fibrous particles with rough-textured surfaces associated with branched structure improves the packable characteristics of these composites during manipulation and molding process [1]. Packable composites have been developed in order to provide resistance to amalgam condensers avoiding slumping and providing tight interproximal contacts by modifying filler size distribution or utilization of fibers [95]. These composites contain high filler contents and have physical and mechanical properties similar to amalgam; however, their clinical performance is more like hybrid composites rather than amalgam and does not present any better properties compared to hybrid ones. These composites are hard to handle, have inconvenient esthetics, and present poor layer match. Condensable composites are used in class II cavity restorations [1, 100].

3.5.4 Curing Modes of Dental Composites

Dental composites used for direct restorations can be activated chemically (self-cure), photochemically (light-cure), or by a combination of both [1].

3.5.4.1 Chemically Activated Resins

Chemically activated resins are available as a base paste and a catalyst paste to mix together. The catalyst paste contains benzoyl peroxide initiator, while the base paste contains an aromatic tertiary amine activator such as N, N-dimethyl-p-toluidine. Mixing these two pastes together at room temperature leads to the formation of free radicals and initiation of additional polymerization. Formation of pores during the mixing process is unavoidable, thereupon polymerization is limited during curing and the structure gets weaker. On the other hand, due to the lack of controllability of the working time of mixed pastes, insertion and contouring steps must be performed as quickly as possible after the mixing. Therefore, chemically activated resins are much more used for restoring buildups that are not easy to be photochemically activated. Formation of oxygen-inhibited layer due to the greater reactivity of oxygen to free radicals compared to monomers prevents the surface polymerization. The advantages and benefits of chemically activated resins are ease of manipulation (working/setting time) by varying proportions and no equipment necessity, hazard-free manipulation, long-term storage stability, and low rates of cross-link formation causing low marginal stress during curing. On the other hand, utilization of these resins presents some shortcomings, which are formation of pores during mixing that weakens the material and increases susceptibility to staining, color instability of aromatic amine accelerators, and difficulty of mixing the components homogenously resulting in unequal polymerization and inconvenient mechanical properties [1].

3.5.4.2 Photochemically Activated Resins

Photochemically activated resins, also called light-cured resins, have been developed in order to overcome shortages of chemically activated resins. Light-cured resins contain photosensitizer and amine initiator, which initiate free radicals using a light source without requiring a mixing process as the previously mentioned resins do. UV-activated composites were the first product of this type that have been replaced by visible blue light-activated composites (VLC). The latter type of light-cured composites provides superior curing depth, and controllable working time. One of the commonly used photosensitizers is Camphorquinone (CQ), at which only 0.2 wt% of it is needed in combination with various amine initiators such as 0.15 wt% of dimethylaminoethyl methacrylate (DMAEMA) [1]. Photosensitive camphorquinone absorbs blue light and gives a slight yellow shade to the composite in its uncured state [3]. CQ is generally accelerated using an aromatic tertiary amine [95]. Photoinitiator system (CQ plus co-initiator (tertiary amine)) concentration is an important parameter affecting the photopolymerization of dimethacrylate polymers. Increased concentration of photoinitiator increases the degree of conversion and hardness of the composite [99]. Other commercial photoinitiators are PPD (1-phenyl-1,2-propanedione), Lucirin TPO (monoacylphosphine oxide), and Irgacure 819 (bisacylphosphine oxide). These photoinitiators are less yellow than CQ [95]. The photopolymerization reaction of dimethacrylate-based monomers consists of three steps, which are (I) initiation (formation of free radicals by irradiation of a light-sensitive initiator), (II) propagation (reaction of free radicals with monomers and opening double bonds of methacrylate groups ($C = C$), which generates a chain reaction resulting in the growth of polymer), and (III) termination (the growth of polymers goes forward till all free radicals are reacted forming covalent bonds). The degree of polymerization or degree of conversion is measured by the percentage of remained double bonds versus primary amount [99]. Polymerization efficiency affects several properties of composites such as degree of conversion, degree of cross-linking, mechanical properties, shrinkage stress, curing depth, trapped free radicals, and biocompatibility. Degree of conversion (DC) is a crucial material property highly related to material properties including mechanical properties, shrinkage, wear resistance, and monomer elution. Degree of cross-linking is another chemical characteristic of composites, which affects mechanical properties and stability of composites. The biocompatibility of resin composites depends on the release of free monomers, release of hydroxyl radicals, and leakage of photoinitiator molecules, which cause mucosal irritation, allergic reactions, and pulp damage. As a matter of fact, optimizing material properties and polymerization is an important subject to be considered. Several factors influence curing and polymerization completion of light-cured resin composites including characteristics of photoinitiator, type and viscosity of monomers, filler content, optical properties, light source, and its irradiation parameters (distance, energy, time of exposure), and temperature [99]. Light-cured composites require no mixing process and make it possible for the operator to perform insertion and contouring prior to curing. Compared to chemically activated composites, light-cured ones need lesser exposure time (about 40 s for a thickness of 2 mm). Besides this, light-cured systems are less sensitive to oxygen inhibition. These composites present several benefits such as low porosity, low staining, good strength, improved color stability due to the absence of aromatic amine accelerator, controlled working time by exposing to an intense blue light, and make it possible to use multiple shades in a restoration owing to the polymerization of small increments at a time. However, these resins still present some drawbacks including cure depth limitation up to 2 mm, sensitivity to room light, requiring special and expensive lamps with hard lamp accessibility in posterior and interproximal regions, long exposure time of dark shades, light absorption of components (ultraviolet absorber color stabilizers, fluorescent dyes, and inhibitors) preventing the completion of reactions, harmful glare to operators, and technique related clinical use [1]. Visible light-cured composites exhibit three serious drawbacks, which are (I) technical limitations of expensive VLC lamps, (II) dependence of curing on light intensity, light angle, and lamp distance forcing to perform incremental placement while using thicknesses greater than 2 mm (due to the limited penetration depth of light) resulting in more time-consuming clinical process of large restorations, and (III) compositional and color limitations of composites. Utilization of LED lamps has solved some of the mentioned problems due to their developed design and operation. These lamps do not release heat, are simpler, lightweight, portable, and are more reliable [1, 101].

Photoinitiator System. Type of the used co-initiator and its ratio to the photoinitiator affects the polymerization progress, biocompatibility, and color of the composite. In spite of the several studies performed on the photoinitiator system, due to the presence of composite variables such as monomer type, filler content, filler morphology, and pigment content, no specific value can be attributed to all composites. Consequently, it is necessary to optimize the concentration ratio of CQ and amine in each case. As CQ results in the yellowish color of dental composites, several photoinitiators have been proposed to replace CQ including phenylpropanedione (PPD), monoacylphosphine oxide (MAPO), bisacylphosphine oxides (BAPO), and benzoyl germanium, which all present higher molar absorptivity than CQ. MAPO is a photoinitiator, which provides higher degrees of conversion in shorter times (about 3 s) compared to those of QC using halogen lamps. Phosphine oxides are type I photoinitiators presenting several benefits such as no need for amines utilization, enhanced biocompatibility, and improved color stability. However, phosphine oxides may cause low curing depths. Briefly, by the side of intrinsic benefits of different photoinitiator systems, extrinsic parameters should be also considered to better choose and balance appropriate composition for each application [99].

Monomers and Fillers Characteristics. The viscosity of monomer system is one of the important characteristics influencing the mobility, reactivity, and DC of dimethacrylate polymers. Fillers also affect the polymerization efficiency by decreasing the DC [99].

Optical Properties. Optical properties of resin composites affect the polymerization of composites. Surface light reflection, light absorption by pigments or photoinitiators, and light scattering by fillers can directly decrease light transmission and curing depth of composites requiring to be optimized [99].

Light Curing Units. Light curing units are one of the extrinsic factors affecting the polymerization of polymers [99]. The four curing lamps used for the photoinitiation of polymerization process are (I) light-emitting diode (LED) lamps, (II) quartz-tungsten-halogen (QTH) lamps, (III) plasma arc curing (PAC) lamps, and (IV) argon laser lamps [99]. Aiming at developing light sources, which provide high conversion associated with low curing stress (using soft-start method in which light intensity gradually increases) various curing lights have been assessed. Halogen and LED lamps are now widely used for dental applications with rather similar powers (about 755 mW/cm^2). As 95% of the spectrum emission of LED lamps ranges between 440 and 500 nm and presents a peak at 465 nm, which is exactly the same absorption peak of camphoroquinone, LED lamps present much more efficiency than halogen lamps when utilized with the camphoroquinone initiator system [100]. LED technology has yielded to a novel family of curing lamps with several benefits such as lightweight, high irradiance, mobility, and long life. However, the light intensity decreases rapidly with distance and the curing areas are not exactly homogeneous because of the placement of several diodes in one light source and heterogeneity of light distribution. The latter can be limited using mixing tubes or diffusing screens. Single-diode blue LED lights are adjusted on the maximum absorption peak of CQ and present better efficiency than halogen lights with broad spectrum in polymerization of CQ-based materials with less time. Utilization of LED lamps with MAPO photoinitiators is less efficient than when it is utilized with CQ and lower curing depth is associated with the use of LED for MAPO composites. Besides this, polymerization is just as important as temperature rise, at which its excessive amount can cause pulpal injuries [99].

Radiant Exposure and Exposure Time. Some scientists believe that flexural strength and curing depth could be increased by increasing radiant exposure and exposure time whereas this theory is still under discussion. In this manner, low molecular weight monomers (e.g. TEGDMA) achieve high DC in conditions with low irradiance (mW/cm^2) and long duration of irradiation (8 min), which is not clinically convenient. As a matter of fact, although theoretically DC of bis-GMA is lower than that of TEGDMA, but through a curing condition at 50 mW/cm^2 with 5 s of exposure, a mixture of Bis-GMA/TEGDMA with a ratio equal to 70/30 presents higher DC than that mixture with a ratio equal to 40/60. However, any changes of irradiation time and irradiance combination can alter the results for different material compositions. So, their selection must be performed precisely [99].

Irradiation Mode. Soft-start curing protocols including ramp, step, or pulse-delay modes are utilized in order to decrease the stress transmission. These protocols retard the onset of polymer gelation, which reduces shrinkage stress. Even though much of studies have been performed on the irradiation modes, there is no certain explanation whether soft-start modes are valuable or not due to the different compositions utilized in each study [99].

Light Distance. Position of light guide is directly attributed to the received energy, which affects the polymerization and DC. This distance should be kept close to the composite. On the other hand, to increase light transmission in the depth of restoration, the light guide tip must be perpendicular to it and should be kept stable [99].

3.5.4.3 Dual-Cured Resins

The development of dual-cured resins has allowed the curing process of composites to be completed and to have increased curing depth by combining the advantages of light and chemical curing systems. Commercial dual-cured resins are available as two photochemically activated pastes in which benzoyl peroxide and aromatic tertiary amine accelerator are present separately. The mixture of two pastes results in a slow self-curing, which can be speeded up on command using related light sources. One still drawback of these composites is the formation of porosity through the mixing process that can be limited using mixing syringes. Besides this, self-cured composites present less color stability compared to that of photo-cured ones due to the presence of amine accelerators in their composition. These composites are suitable for luting ceramic inlays where light penetration is limited [95].

3.5.5 Longevity and Failure of Dental Composites

There is a growing body of systematic reviews and meta-analysis evaluating longevity, survival rates, and main reasons for failure of resin-based composites with compositions produced by different manufacturers applied in different clinical cases. It has been demonstrated in several studies that survival rates of dental composites ranging between 55% and 95% during 5 years for posterior composites restorations, which is two to three times lower than that of amalgam restorations [105]. Opdam et al. have discussed in their review study survival rates of posterior resin composites. Their evaluation demonstrated that resin composites exhibit good survival rates presenting annual failure rates (AFR) of 1.8% and 2.4% after 5 and 10 years of restoration placement, respectively [106]. Nieuwenhuysen et al. have reported the longevity of composites to be about 7.8 years, whereas it increases to 12.8 years for amalgam restoratives. Other studies have found rather similar survival rates for these two restoratives [99]. For example, Ferracane et al. have concluded that the longevity of composites presents very low (about 2%) annual failures through 10–20 years, which is slightly shorter than amalgams [95]. Several studies were performed regarding to determine the major sources of composite failure. Evaluations have shown that secondary caries, fracture, marginal deficiencies, and wear are the most probable reasons of composites failure [1, 95]. In this manner, secondary caries and restoration fracture seemed to be the most important reasons of composite failure [105]. At the time, tooth fracture has been found to be the most usual reason of failure for amalgams [96]. Generally, in the first year of composite placement endodontic complications are the main failure reasons, whereas afterward caries and fractures are the main reasons of failure. On the other hand, failure risk of molar restorations is higher than that for premolar restorations [106]. Secondary caries are formed due to the polymerization shrinkage and related stress development affecting the interfacial bonds; whereas, fracture is formed due to the poor mechanical properties of the composite material. Layering placement of dental composites for restoring cavities ensures the completion of curing process and decreases polymerization shrinkage stress [95]. Studies have shown that conventional dental composites are more prone to failures than hybrid composites with larger particles [105]. Larger composite restorations are more prone to secondary caries in patients with high-caries risk [99]. On the other hand, some studies have shown that when large restoratives are in demand, composites present longer longevity than amalgam in cases with low risk of caries formation whereas in cases with high risk of caries the two restoratives present the same survival rates [95]. Choosing between direct or indirect methods for restoring posterior teeth depends on the cavity size. In this manner, small or medium cavities are generally restored with direct resin composites whereas large cavities are restored via indirect restorations. Systematic reviews and meta-analysis evaluations regarding the differences of direct and indirect resin composites in their clinical performance as permanent posterior restorations have demonstrated that there is no evident difference in longevity of these two restorations. Due to the fact that direct resin composites are easier to be manipulated, and are inserted in just one session, they are widely used in restorative dentistry even in large cavities [107]. Table 3.35 lists the main reasons for composite restoration failures presented in several studies over different duration times.

3.5.5.1 Polymerization Shrinkage Stress

Stress developed through resin composites generated from polymerization shrinkage is a multi-factorial process creating some adverse effects on dental restorations such as interfacial defects and generation of contraction stress due to the employed constraint by cavity walls. Contraction stress is also affected by viscoelastic behavior

Table 3.35 Main reasons of composite restoration failure presented in different studies over different duration times (Reprinted from Clinical Oral Investigations, Brunthaler et al., Longevity of direct resin composite restorations in posterior teeth: a review, no. 68, Copyright (2003), with permission from Springer) [105]

Reason of failure	0–5 years	6–17 years	0–17 years
Secondary caries	2	5	7
Fracture of restoration	3	0	3
Marginal defect	1	0	1
Hypersensitivity	1	0	1
Periapical abscess	1	0	1
Loss of anatomic form	0	1	1

and elastic modulus of the material. Several studies have proven the presence of a directly related correspondence between polymerization shrinkage and marginal integrity in dental restorations of class V. Volumetric shrinkage of composites is a consequence of the formation of covalent bonds between monomers. The amount of this shrinkage in composites is related to filler content, composition, and DC of the resin matrix. Volumetric shrinkage of Bis-GMA and TEGDMA is equal to 5.2% and 12.5%, respectively, which are higher than those of typical composites (about 2–3%) with high filler contents such as hybrid and microfilled composite. In photo-activated materials this value is directly related to their DC. The rapid reaction generated in these composites hinders viscous flow providing rigidity of the polymeric matrix in a few seconds with low DC. Consequently, stress development starts after polymerization initiation and it increases slowly within the time. Light-cured composites present higher contraction stress than self-cured composites due to the slow reaction rate and formation of fewer free radicals within the self-cured composites with lower DC [98].

Volumetric shrinkage is not the sole indicator of composite stress development because composites possessing with high amounts of inorganic fillers exhibit high elastic modulus as well that can increase stress within the composite. Lower elastic modulus of microfilled composites brings about lower stress amounts compared to hybrid composites. On the other hand, studies have shown that increase of the viscous flow of composites reduces the stress development. Viscoelastic behavior of resin composites is complex dependent on the filler content of composite, matrix composition, and DC. Applying an intermediary layer with low elastic modulus increases the compliance of the cavity walls and causes stress relief [98].

"Soft start" is a photo activation method based on the utilization of small light irradiance through the initial seconds of light activation, and is used to decrease the polymerization rate as well as to increase viscous flow duration. Several studies have proven the benefits of this method in reducing volumetric shrinkage. However, decrease of reaction rate does not always provide reduction of contraction stress since in some materials such as dimethacrylate composites elastic modulus develops at low DC. Decreasing the contraction rate by modifying the polymerization rate has some shortcomings such as generation of alterations within the polymer, susceptibility to ethanol degradation, and low cross-linking density. On the other hand, presence of diluent monomers in the composition of composites also increases the volumetric shrinkage and contraction stress as well as DC. To conclude, degree of conversion and volumetric shrinkage are the major reasons generating contraction stress within dental restorative composites. Actually, contraction stress of regular or packable composites is highly related to filler content and is rather independent of the polymer composition [98].

3.5.5.2 Wear of Dental Resin Composites

Dental composites, which are currently used in dental applications present convenient properties for being used in almost all oral areas; however, there are still some concerns regarding their use in high stress regions. The issue whether wear of dental composites is still a matter of concern has been assessed in many studies [95].

Systematic reviews have shown that wear mechanisms are one of the several reasons leading to fracture of resin composites in posterior regions where they are subjected to mechanical forces and chemical substances. The amount of wear in materials is related to three terms, which are (I) the structure of the material (type of material and its geometry), (II) interaction conditions (applied forces and their duration), and (III) the environment and surface conditions (temperature, surface geometry, and its chemistry). Mechanism of wear can be classified into four groups including adhesive, abrasive, fatigue, and corrosive wear. The field of dentistry wear is generally classified into three terms, which are (I) attrition (wear at contact regions), (II) abrasion (wear at non-contact regions), and (III) corrosion (wear caused by chemicals) [108].

Due to the presence of saliva in the oral cavity, which acts as a lubricant, adhesive wear of resin composite restorations is limited. In the oral environment, abrasion wear is a crucial mechanism, which is dependent on the emplacement of restoration. Shape, size, orientation, amount, and distribution of filler particles within the resin composite as well as the type of resin matrix have a crucial effect on the abrasion rate of these composites. Moreover, the quality of bonding between filler particles and resin matrix plays an important role in the rate of abrasion wear of resin composites. Studies have shown that increase of size or filler content, increased DC, and enhanced bonding quality between fillers and resin matrix decrease abrasive wear. Abfraction is a type of fatigue wear occurring in tooth tissue due to the repeated cycle of mastication. Corrosive wear is another wear mechanism affecting tooth structure by its exposure to corrosive chemicals, which can soften and roughen resin composites increasing the risk of abrasive wear [108].

Previous studies have demonstrated that higher filler contents largely affect the wear resistance of conventional (high viscosity) resin composites by decreasing its level. Despite this, the evidence is not more applicable due to the fact that novel composites have large filler volumes and the observed variations of wear are not simply related to the filler content. Some studies have shown that nanofilled composites present higher wear resistance compared to conventional resin composites; however, few researchers have evaluated the wear rates of these novel composites. Studies have shown that generally flowable resin composites applied in posterior regions demonstrate similar wear rates to that of nanofilled composites and incorporation of spherical submicron filler particles or dense nanoparticles may enhance the wear resistance of flowable resin composites [108].

To conclude, the wear development of restoratives used in typical cases seems to be the least important issue of new composite materials; however, fracture and wear are crucial factors for patients with bruxing and clenching or parafunctional habits. Furthermore, wear of large composite restorations in direct occlusal contact is still a clinical concern [95, 108].

3.5.6 Laboratory Composites

Laboratory composites are indirect restorative composites mostly used for large cavities to fabricate prosthetics extra-orally. These composites are also used for fabricating crowns, inlays, veneers, and coronal restoration on dental implants where a large amount of masticatory force is applied to the restoration. In cases with large cavities where the risk of polymerization stress, microleakage, postoperative sensitivity, poor interproximal margins, and recurrent caries are serious, indirect restoratives are the best clinical choice. The composition of these composites is similar to that of direct composites and are generally bonded using resin cements [3].

3.5.7 Core Build-Up Composites

Core build-up composites are used in cases where a large amount of dentin structure has been injured and lost but there is still a sufficient tooth structure able to hold and preserve the restoration. These composites typically are colored to provide distinction from tooth structure. They have some advantages over amalgams such as better esthetics, ability to form bonds with dentin, and rapid finishing [3].

3.5.8 Low-/Non-Shrinkage Composites

Shrinkage of composites occurs during the curing process. The level of shrinkage is rather small (typically 3–3.5 vol.%) but rapid (5–10 s) and it can largely affect various properties of the composite by inserting high stress levels to healthy tooth structure or formation of porosities along filler interfaces, which increases the risk of failure and water absorption. The level of shrinkage can be controlled by adjusting filler portion or monomer composition and mixture. The first low-shrinkage composites were introduced by 3 M and ESPE in 2000 based on epoxy ring-opening reactions [101].

Utilization of resins with high molecular weights reduces the curing shrinkage and associated stress. Hydrogenated dimer acids admixed with bis-GMA and UDMA are examples of monomers with higher molecular weights, higher degrees of conversion, lower double bond concentration, lower water adsorption, and lower shrinkage compared to those of dimethacrylate resins [99, 102]. Additionally, the low cross-link density of hydrogenated dimer acids provides polymers with great flexibility and low elastic modulus. Commercial nanohybrid resin-based composites, which are based on dimer acid (N'Durance, Septodont, France) contain Bis-GMA, UDMA, and dicarbamate within their polymeric matrix and present appropriate flexural strength and elastic modulus [102]. The use of ring-opening systems such as oxirane-based resins has been developed in dental composites. Oxirane resins provide several advantages over Bis-GMA-based dental composites such as increased depth of cure, higher strength, and lower polymerization shrinkage; however, some researchers believe in the cytotoxicity of oxirane resins. As a matter of fact, several studies have focused on methods reducing the cytotoxicity of ring-opening monomers. Silorane generated from the reaction between oxirane and siloxane molecules was first introduced by Weinmann. Silorane systems (Filtek Silorane; 3 M-ESPE) present low volumetric shrinkage (due to the presence of ring-opening oxirane), high hydrophobicity (due to the presence of siloxane), and convenient mechanical properties. Further studies have shown that siloranes are chemically stable in oral fluids and can be successfully used for dental restoratives [102]. Other monomers such as urethane dimethacrylate resin DX511 have been also used to decrease composite shrinkage owing to its higher molecular weight than bis-GMA and UDMA. Additionally, TCD-DI-HEA, a urethane monomer of Venus Diamond (Kulzer) having low viscosity provides not only low shrinkages but also low polymerization contraction stress [95].

3.5.9 Antibacterial Composites

Resin composites are the restorative materials, which tend to accumulate more biofilms than other dental restoratives such as amalgam and ceramics. Biofilms produce carbohydrates and acids, which increases the risk of composite degradation in the oral cavity and secondary caries that is the most important reason of resin composite failure [74]. As a matter of fact, developing resin composites with antibacterial and antifungal activities can lead to a great advance in enhancing lifetime of these composites. Consequently, polymerizable quaternary ammonium methacrylates (QAMs) and antibacterial fillers are added to the composition of dental composites. 12-methacryloyloxydodecyl-pyridinium bromide (MDPB) is an example of antibacterial monomer used in dental composites, which hinders formation of biofilms around composites without any negative influence on the mechanical properties or polymerization conversion of resin composites. Besides this, silver, zinc oxide, and bioglass are examples of frequently used antibacterial fillers, which release antibacterial ions through time. Incorporation of quaternary ammonium polyethylenimine (QPEI) nanoparticles into a resin composite has also demonstrated significant anti-biofilm properties [109]. Quaternary ammonium compounds (QACs) are materials commonly used in the composition of resin composites restoratives as monomer and micro- or nanofiller. QACs, with chain length from about 12 to 16 present significant antimicrobial activity in resin composites. Recent studies have shown that immobilized antibacterial agents such as QACs present the best long-term antibacterial activity; however, excess amounts of QAMs may bring cytotoxicity to living cells and affect the network structure, degree of conversion, solvent sorption, polymerization shrinkage, and mechanical properties of resin composites. Consequently, the proper amount of these antimicrobial agents should be optimized [110]. Nowadays, dental resin composites are generally bioinert and the combination of both antibacterial fillers and monomers within a resin composite is the future trend in composites development in order to enhance their long-term performance [109].

3.5.10 Fiber-Reinforced Composites (FRCs)

Fiber-reinforced composites (FRCs) are a new generation of nonmetallic dental materials used in restorative dentistry since 1960 s. FRCs are composed of three main components, which are the polymeric matrix, reinforcing fibers, and the interphase region [111]. FRCs have been largely modified through the years with the aim of better mixing the polymeric material with reinforcing fibers and enhance its handling properties. Reinforcing fibers can present four different structures in the resin matrix, which are continuous unidirectional (roving), continuous bidirectional (weaves), continuous random-oriented (mat), and short random-oriented [96]. Glass fibers are widely used in this composition due to their transparency, good surface chemistry, and excellent adhesion to the polymeric matrix owing to the presence of hydroxyl groups, which react with silane coupling agents [111]. Glass fiber reinforced composites (GFRCs) in dental applications are consisting of amorphous, homogeneous, and three-dimensional structure of glass fibers, which are embedded in polycarbonate, polyurethane, or acryl-based polymers (PMMA, Bis-GMA). Silane coupling agents acting on the surface of glass fibers provide chemical bonds between glass fibers and polymeric matrix. Table 3.36 presents the composition of commercial dental FRCs [112]. The main glass fibers used in fiber-reinforced composites are continuous unidirectional E-glass fibers, which are chemically stable in pH ranges of 4–11 [96]. Glass E is a calcium–aluminum–borosilicate glass containing low concentrations of alkalines with high electrical resistance and chemical stability. Incorporation of other fibers including carbon/epoxy, polyaramide, and ultrahigh molecular-weight polyethylene (UHMWP) have also been assessed; however, inappropriate color of carbon/epoxy fibers, poor polishability, and handling properties of polyaramide, and low adhesion of UHMWPE fibers to the matrix has greatly restricted their application in dentistry. Herein, glass fibers with excellent adhesion to polymeric matrix and high mechanical properties as well as superior esthetics have gained much attention [112].

Factors influencing the overall physical and mechanical properties of FRCs are impregnation of fibers with resin, bond quality between fibers and matrix, fiber volume fraction, and fiber orientation. Different orientations of fibers provide different properties to FRCs, for example, for fixed partial dentures and root canal post in which high strength and rigidity are required, continuous unidirectional fibers are appropriate. On the other hand, for restoring crown margins it is more convenient to use bidirectional fibers, which provide greater toughness. Increasing fiber quantity (to about 70 vol.%), complete fiber impregnation with resin matrix, and continuous unidirectional orientation of fibers provide high flexural strength and elastic modulus to FRCs [96].

Nowadays, developed FRCs are widely used in various dentistry applications including removable dentures, direct or indirect pontics, endodontics (prefabricated or customized root canal anchoring systems), conservative dentistry (direct and indirect fillings, inlays, and onlays), orthodontics (retention splints, space maintainers, active "en-masse" units, metal-free brackets, and orthodontic wires), periodontology (periodontal splints and posttraumatic splints), restoring veneers, and pediatric dentistry (crowns in primary molars, splints, space maintainers, and direct fillings). The benefits of FRC applications in dentistry over other dental restoratives are their excellent esthetics, great biocompatibility, easy manipulation, minimally invasive approach, minimal laboratory work (usually frameworks can be prepared at chairside, directly in the oral cavity), and high mechanical properties. The main drawbacks and disadvantages of these composites are the lack of information about

Table 3.36 Composition of commercial glass fibers (Reprinted from Materials Science and Engineering: C, Khan et al., An update on glass fiber dental restorative composites: a systematic review, no. 28, Copyright (2015), with permission from Elsevier) [112]

Core of glass fiber-reinforced composite	Manufacturer	Composition
Preimpregnated E-glass FRC	Vectris Pontic, Ivoclar Vivadent, Schaan, Liechtenstein	Bis-GMA (24.5%) Triethyleneglycol dimethacrylate (6.2%) Decandiol dimethacrylate (0.3%) Urethane dimethacrylate (0.1%) Highly dispersed silica (3.5%) Catalysts and stabilizers (0.3%) Pigments (0.1%), Preimpregnated E-glass fibers (65.0%)
Preimpregnated S-glass FRC	FiberKor, Pentron Corporation, Wallingford, CT, USA	Preimpregnated S-glass fibers ($\simeq$ 60%) in a 100% Bis-GMA matrix
Impregnated E-glass FRC	Stick StickTech, Turku, Finland	E-glass fibers impregnated with poly(methyl methacrylate)

their clinical longevity, hydrolysis, and degradation of the interface between fibers and organic matrix, and their high cost [96, 111].

3.5.11 Polymer Infiltrated Ceramic Networks (PICNs)

Due to the poor mechanical properties and wear resistance of the first machinable resin composite blocks (ParadigmTM, 3MTM, St Paul, USA), polymer infiltrated ceramic networks (PICNs) were developed by Vita (VITA Zahnfabrik, Bad Säckingen, Germany) to overcome these problems. These materials are resin-ceramic composites composed of two interconnected networks, which are ceramic matrix and infiltrated polymer. The manufacturing technology of these composites is similar to glass-infiltrated ceramics; however, the final shrinkage of polymers (about 5%) is greater than that of glass (about 1%). PICNs have elastic modulus rather close to that of dentin, and are milled or repaired easily. Enamic (Vita, Bad Sachingen, Germany) is a PICN product introduced in 2013 based on infiltration of a monomer mixture into a partially sintered ceramic. The usually incorporated polymers are urethane dimethacrylate (UDMA) and triethylene glycol dimethacrylate (TEGDMA). Ceramic network is generally composed of SiO_2 (58–63 wt%), Al_2O_3 (20–23 wt%), Na_2O (9–11 wt%), K_2O (4–6 wt%), B_2O_3 (0.5–2 wt%), CaO (<1 wt%), and TiO_2 (<1 wt%). Appropriate elastic modulus (about 30 GPa) and strength (about 135 MPa) of PICNs have made them a good choice for restoring crowns, inlays, onlays, and veneers. Due to the curing shrinkage of resin materials within the ceramic network, interfacial gaps form between the ceramic and the polymer, which opacifies the structure. Owing to the low elastic modulus of these materials, they are successfully applied for prosthetics on stiff implants [17].

3.5.12 Compomers

Compomers also called "resinomers" and "giomers" are polyacid-modified composite resins [34]. The name is derived from two terms "comp" and "omer" from composite and ionomer, respectively, combining both advantages of these two materials. These polyacid-modified composite resins are a type of esthetic fluoride-releasing materials introduced in the 1990 s. Compomers contain polyacid-modified monomers with carboxylic acid functional groups such as diester of 2-hydroxyl methacrylate, Bis-GMA, or TEGDMA and fluoride-releasing silicate glass particles (GIC) as well as an initiator system and coupling agent [113]. These water-free composites are filled with 42% to 67% filler particles (nonreactive inorganic filler particles and reactive silicate glass particles) with particle sizes ranging between 0.8 μm and 5.0 μm. The setting reaction of these one-component compomers is similar to traditional resin-based composites in which addition polymerization occurs via light-initiated polymerization of the acidic monomer. Chemically activated and dual-cured materials are also utilized in some powder-liquid compomers in order to increase reaction rate; however, light-initiated systems are much more usual. The initiator system is generally camphorquinone accompanied with amine accelerator. Although compomers are hydrophobic materials, setting reaction is further continued by a slow acid–base reaction between glass fillers and the acid groups of functional monomers by their placement in the vicinity of saliva, which leads to fluoride release. Accordingly, as compomers are sensitive to water or moisture, they should be packed thoroughly [1, 3].

Compomers absorb up to 3.5 wt% water, which gradually neutralizes the carboxylic acid groups. Moisture uptake of compomers facilitates fluoride release from the glass filler to the resin matrix to be further released into the mouth. Acidic and basic conditions increase the fluoride release. Studies have shown that the curing depth of compomers is rather similar to that of resin composites and is dependent on the manufacturer, color shade, and composition. The value of polymerization contraction stress is also related to the manufacture varying approximately between 1.27 MPa and 3.41 MPa for dyrcat and compomer F2000 (3 M), respectively, which are similar to that of hybrid composites. As rapid curing with high values of polymerization contraction stress could result in restoration or bond failure, it seems that Dyract compomer provides better properties as a restoring material. In general, compressive strength, biaxial flexural strength, diametral tensile strength, and surface hardness of compomers are comparable to those of conventional resin-based composites whereas the fracture toughness of compomers (0.97–1.23MPam$^{1.2}$) is slightly lower than that of composites (1.75–1.92MPam$^{1.2}$). The low fracture toughness of compomers is the reason why these materials are not ideal restoratives for high stress-bearing regions. On the other hand, some studies have shown that water uptake of compomers leads to a decrease of the compressive and flexural strength through the time even up to 40% after a couple of weeks; however, this effect does not demonstrate considerable impact on the clinical performance of these materials [113]. It should be mentioned that studies have shown that surface hardness and micro-tensile strength of compomers do not change in saliva solution and surface treatment of reactive glass particles with silane agent, reduces water uptake and minimizes strength variation [95].

These composite materials should be bonded to the tooth structure via bespoke luting agents [113]. In order to give self-adhesive properties to water-free compomers, a dentin-bonding agent should be used before their placement. The strength of the formed bond is equal or higher than that of hybrid composites. Contrary to the water-free compomers, powder-liquid compomer cements with some water content are self-adhesive due to the acidity given by water to the liquid [95].

Compomers are practically applied for the same clinical applications as resin-based composites such as restoring low stress-bearing regions (class I, class II, class III, and class V cavities), fissure sealants, cementing orthodontic bands, and paedodontics for children's dentistry. Until now, compomers have shown successful and acceptable clinical performance as well as conventional composite resins through their clinical applications. However, some studies have shown color instability and marginal discoloration of compomers due to their water absorption from coffee or red wine, which cause discoloration of compomer surface and stains [95, 113].

3.5.13 Functionally Graded Materials (FGMs)

Functionally graded materials (FGMs) are a kind of modern composite materials formed of two or more phases with compositional, microstructural, or porosity gradient according to a predetermined pattern in order to provide varying functions and properties. Bone, teeth, human skin, wood, and bamboo are examples of natural FGMs. Tooth is an FGM in a manner that the wear resistance of the tooth structure slightly decreases from its external surface (enamel) to its internal core (dentine), which is more ductile in order to minimize shock incomes. Among the different types of FGM, porosity gradient functionally gradient materials are used in dental applications [114].

The main failure reason of composite materials is their delamination from the sharp interfaces separating two types of materials in which high stress concentration occurs. Gradient interfaces within the FGMs solve this problem by gradually introducing the second phase of material into the first one. The first introduction of FGM was carried out in the early 1980 s in Japan. The material was utilized as a thermal barrier in aerospace applications. The idea of FGM has brought several benefits in material performance including enhanced elastic modulus, tensile strength, and wear resistance. These materials have been largely studied for dental restorations and other biomedical applications [114].

The fabrication of FGM consists of two main steps including producing the homogeneous-graded structure and the consolidation of the graded materials. Physical and chemical vapor deposition, self-propagating high-temperature synthesis (SHS), powder metallurgy, centrifugal casting, slip casting, tape casting, and additive manufacturing are methods used for fabricating FGMs [114].

Functionally graded materials are used as composite dental posts in order to retain the core of restored tooth with minimized harm to root structure by gradually decreasing the stiffness of the paralleled post with high coronal stiffness and low apical stiffness. This pattern of stiffness variation minimizes stress concentration and interfacial shear stress by gradually transferring the applied stress from the core and post to the dentine [115]. FGM is also used in porcelain-fused-to-metal restorations (PFM) in which the metal-ceramic bond has been improved. Applying functionally graded microstructure improves the mechanical and thermal properties of materials by eliminating effects of sharp interfaces. FGMs are utilized in various fields such as thermal barrier coatings, biomechanics, mechanics, power industry, automotive industry, aerospace, civil, nuclear, aeronautic, and naval engineering. Ceramic FGMs are used for implant abutments due to their high fracture toughness, low modulus of elasticity, excellent esthetics, and corrosion resistance. Within the last few years, FGMs have been developed for dental implants coated with hydroxyl apatite (HA), Zr, Ti, and their oxides in order to minimize mechanical and biological mismatch between natural structures and implant and provide enhanced biocompatibility and longevity. FGMs are produced via two methods, which are constructive processes (producing gradients by selectively stacking two or more materials) and transport-based processes (by utilizing natural transport phenomena), which are further classified into gas-based, liquid-phase, solid-phase, and biopolymeric-based processes. Examples of gas-based processes are chemical vapor deposition (CVD), physical vapor deposition (PVD), ion plating, plasma spraying, and ion mixing. Centrifugal casting, combustion, tape casting, slip casting, gel-casting, electrophoretic deposition, chemical solution deposition, laser deposition, and directional solidification are examples of liquid-phase processes. Besides these, solid-phase processes include spark plasma sintering (SPS), solid free-form fabrication (SFF), laser cladding–based method, selective laser sintering (SLS), selective laser melting (SLM), three-dimensional (3D) printing, and powder metallurgy (PM). On the other hand, freeze-drying, compression molding, and electrospinning are based on biopolymeric processes [116].

3.5.14 Self-Adhesive Restorative Composites (SACs)

Self-adhesive (flowable) composite restoration technology is a minimal invasive method applied to restore injured teeth without any need to acid etch the restorative composite providing simpler and faster application. These composites bond to the tooth enamel and dentin directly without any need for any other adhesive or luting material [117]. Self-adhesive restorative composites (SACs) have been commercially available since 2009 for dentistry applications. These composites have low viscosities and are used for small class I and non-carious cervical lesions. Fusio Liquid Dentin (Pentron Clinical, Orange, CA, USA), Vertise Flow (Kerr, Orange, CA, USA), and Embrace Wetbond (Pulpdent, Watertown, MA, USA) are three self-adhesive composite products available in the market. The main resin components of these products are aliphatic diurethane dimethacrylate (UEDMA), triethylene glycol dimethacrylate (TEGDMA), hydroxyethyl methacrylate (HEMA), trimelitic acid methacrylate (4-MET), glycerolphosphoric acid dimethacrylate (GPDMA), bisphenol glycidil dimethacrylate (Bis-GMA), bis-methacryloyloxy ethyl phosphate (BMEP), and trimethyloyl propane trimethacrylate (TMPTMA). Typical fillers used in the composition of self-adhesive composites are silanated barium glass, NaF, nano-sized colloidal SiO_2, and YF_3. These composites present inferior bond strength to teeth compared to conventional composites [97]. Almost all of the publications are carried out on Vertise Flow composite [118]. Vertise Flow is consisting of phosphoric acid ester methacrylate and glycerolphosphate dimethacrylate (GPDM), which presents inferior microleakage to conventional fissure sealants as well as good marginal sealing and low polymerization stress [97, 117]. Furthermore, pre-etching of enamel and dentin with phosphoric acid enhances bonding quality and marginal sealing of Vertise Flow by providing micro-retention. A comparison performed on bonding strength and its effectiveness on the two Fusio Liquid Dentin and Vertise

Flow self-adhesive composites demonstrated that Fusio Liquid Dentin presents better results than latter due to the composition difference and type of functional monomer. Fusio Liquid Dentin consists of 4-methacryloxyethyltrimetellitic acid (4-MET) as the functional monomer, which can chemically bond to hydroxyapatite (HAp) and tooth tissue. The functional monomer of glycerol phosphate dimethacrylate (GPDM) presented in the structure of Vertise Flow etches HAp instead of creating bonds. On the other hand, Fusio Liquid Dentin is more fluid and presents better wettability than Vertise Flow owing to its lower amounts of filler, which facilitates its handling and emplacement [118]. The retention rate and bond strength of self-adhesive composites are less than 3-step etch-and-rinse, 2-step etch-and-rinse, 2-step self-etch, and 1-step self-etch adhesives. In fact, adhesives that are applied in multi-steps provide stronger bonds, better micro-tensile bond strength, higher and interfacial fracture toughness than those of one-step adhesives [117]. These composites have not been thoroughly examined and there is a lack of information about their physical, mechanical, or optical properties, which seem to be dependent on the composition of these composites [97].

3.5.15 Self-Healing Composites (SHCs)

Initiation and growth of microcracks is the major reason of composite failure, which can be minimized by crack healing strategies. Crack healing technology is a great step forward in polymer science [97]. A self-healing dental composite (SHDC) consists of four major components, which are the polymer matrix, reinforcing fillers, healing powder (HP) such as strontium fluoroaluminosilicate particles, and healing liquid (HL) such as aqueous solutions of polyacrylic acids encapsulated in silica microcapsules, which protects HL during composite preparation. The elastic modulus of silica microcapsules perfectly matches the elastic modulus of fillers, which increases its ability to protect HL while composite preparation. Moreover, due to the acidity and hydrophilic properties of HL, silica is the best option for encapsulating the HL [119]. White et al. for the first time formulated a self-healing dental composite consisting of about 40 wt% Bis-GMA/UDMA/TEGDMA resin component, 55 wt% silanated silica, 5 wt% DCPD-filled UF microcapsules, and 2 wt% Grubbs' catalyst. This composite presented an average of 57% healing efficiency after 7 days at room temperature. Compomers containing silanated silica microcapsules (filled with diluted polyacrylic acids) and fluoroaluminasilicate glass particles incorporated in a photocurable Bis-GMA/HEMA matrix are the most recently introduced self-healing composites [97]. The mechanism of self-healing process in dental resin composites includes the releasement of HL from microcapsules during cracking followed by its reaction with HP resulting in the formation of GIC in the crack, which acts as a defect healer (Fig. 3.43). Formation of GIC provides many outstanding properties to the composite owing to its biocompatibility, adhesion to tooth, low shrinkage, and fluoride releasement. Functionality of self-adhesive dental composites can be enhanced

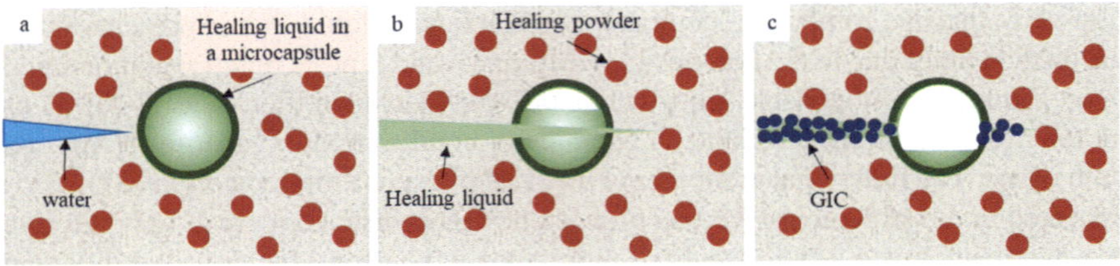

Fig. 3.43 Self-healing mechanism of a SHDC. **a** Formation of cracks containing water, **b** microcapsule failure and releasement of HL, **c** reaction of HL with HP resulting in the formation of GIC. Redrawn from Huyang et al. [30]

by adding antimicrobial agents or replacing a portion of HP with hydroxyapatite, which increases the remineralization of tooth [30].

3.5.16 Remineralizing Composites

Remineralizing composites favor the releasement of ions (e.g. calcium and phosphate), which facilitates the deposition of hydroxyapatite (HAP) crystals on the surface of teeth (enamel). Calcium orthophosphate (CaP) particles are an example of ion-releasing fillers incorporated in the composition of resin-based composites. Several studies have been performed on the efficiency of different phases of CaP. A study performed on a composite consisting of 40 wt% amorphous calcium phosphate (ACP) embedded in a Bis-GMA/TEGDMA/HEMA resin matrix presented 71% efficiency for recovering mineral content of non-cavitated bovine enamel lesions after 28 days. The remineralizing effect of CaP containing composites has been conducted in a few numbers of studies and there is still a lack of literature and clinical information regarding to efficiency, longevity, and long-term performance of these composites within oral environment [97].

References

1. Anusavice, K. J., Shen, C., & Rawls, H. R. (Eds.). (2012). *Phillips' science of dental materials*. Elsevier Health Sciences.
2. Höland, W., Schweiger, M., Rheinberger, V. M., & Kappert, H. (2009). Bioceramics and their application for dental restoration. *Advances in Applied Ceramics, 108*(6), 373–380.
3. Powers, J. M., Sakaguchi, R. L., & Craig, R. G. (2012). *Craig's restorative dental materials/edited by Ronald L. Sakaguchi, John M. Powers*. Philadelphia, PA: Elsevier/Mosby.
4. Bergmann, C. P., & Stumpf, A. (2013). *Dental ceramics microstructure, properties and degradation, topics in mining, Metallurgy and Materials Engineering*. Berlin, Heidelberg: Springer.
5. Zhang, Y., & Kelly, J. R. (2017). Dental ceramics for restoration and metal veneering. *Dental Clinics of North America, 61*(4), 797–819.

6. Zanelli, C., Raimondo, M., Guarini, G., & Dondi, M. (2011). The vitreous phase of porcelain stoneware: composition, evolution during sintering and physical properties. *Journal of Non-Crystalline Solids, 357*(16–17), 3251–3260.
7. Saint-Jean, S. J. (2014). Dental glasses and glass-ceramics. In *Advanced Ceramics for Dentistry* (pp. 255–277). Butterworth-Heinemann.
8. Ho, G. W., & Matinlinna, J. P. (2011). Insights on ceramics as dental materials. Part I: ceramic material types in dentistry. *Silicon, 3*(3), 109–115.
9. McLaren, E. A., & Cao, P. T. (2009). Ceramics in dentistry—part I: classes of materials. *Inside dentistry, 5*(9), 94–103.
10. Kelly, J. R. (2008). Dental ceramics: What is this stuff anyway? *The Journal of the American Dental Association, 139,* S4–S7.
11. Shenoy, A., & Shenoy, N. (2010). Dental ceramics: An update. *Journal of conservative dentistry: JCD, 13*(4), 195.
12. Denry, I., & Holloway, J. (2010). Ceramics for dental applications: a review. *Materials, 3*(1), 351–368.
13. Pollington, S., & van Noort, R. (2009). An update of ceramics in dentistry. *Int J Clin Dent, 2*(4), 283–307.
14. Höland, W., Schweiger, M., Watzke, R., Peschke, A., & Kappert, H. (2008). Ceramics as biomaterials for dental restoration. *Expert Review of Medical Devices, 5*(6), 729–745.
15. Conrad, H. J., Seong, W. J., & Pesun, I. J. (2007). Current ceramic materials and systems with clinical recommendations: a systematic review. *The Journal of prosthetic dentistry, 98*(5), 389–404.
16. Kelly, J. R., & Benetti, P. (2011). Ceramic materials in dentistry: historical evolution and current practice. *Australian Dental Journal, 56,* 84–96.
17. Silva, L. H. D., Miranda, R. B. D. P., Favero, S. S., Lohbauer, U., & Cesar, P. F. (2017). Dental ceramics: A review of new materials and processing methods. *Brazilian oral research, 31.*
18. Höland, W., Rheinberger, V., Apel, E., van't Hoen, C., Höland, M., Dommann, A., … & Graf-Hausner, U. (2006). Clinical applications of glass-ceramics in dentistry. *Journal of Materials Science: Materials in Medicine, 17*(11), 1037–1042.
19. Denry, I., & Kelly, J. R. (2014). Emerging ceramic-based materials for dentistry. *Journal of Dental Research, 93*(12), 1235–1242.
20. de Carvalho Ramos, N., Campos, T. M. B., de La Paz, I. S., Machado, J. P. B., Bottino, M. A., Cesar, P. F., et al. (2016). Microstructure characterization and SCG of newly engineered dental ceramics. *Dental Materials, 32*(7), 870–878.
21. Raigrodski, A. J. (2004). Contemporary materials and technologies for all-ceramic fixed partial dentures: a review of the literature. *The Journal of Prosthetic Dentistry, 92*(6), 557–562.
22. Denry, I., & Kelly, J. R. (2008). State of the art of zirconia for dental applications. *Dental Materials, 24*(3), 299–307.
23. Turon-Vinas, M., & Anglada, M. (2018). Strength and fracture toughness of zirconia dental ceramics. *Dental Materials, 34*(3), 365–375.
24. Piconi, C., Condo, S. G., & Kosmač, T. (2014). Alumina-and zirconia-based ceramics for load-bearing applications. In *Advanced ceramics for dentistry* (pp. 219–253). Butterworth-Heinemann.
25. Yin, L., Song, X. F., Song, Y. L., Huang, T., & Li, J. (2006). An overview of in vitro abrasive finishing & CAD/CAM of bioceramics in restorative dentistry. *International Journal of Machine Tools and Manufacture, 46*(9), 1013–1026.
26. Hill, E. E. (2007). Dental cements for definitive luting: a review and practical clinical considerations. *Dental Clinics of North America, 51*(3), 643–658.
27. Wilson, A. D., & Nicholson, J. W. (2005). *Acid-base cements: their biomedical and industrial applications.* Cambridge University Press.
28. Sunico-Segarra, M., & Segarra, A. (2015). *A practical clinical guide to resin cements.* Berlin Heidelberg: Springer.
29. Albers, H. F. (2002). *Tooth-colored restoratives: principles and techniques.* PMPH-USA.

30. Kumar, M., & Kumari, S. (2016). Resin-modified Glass Ionomer Cements and its Use in Orthodontics-Concept Old is Gold: View Point. *International Journal of Dental and Medical Speciality, 3*(3), 10.
31. Wingo, K. (2018). A review of dental cements. *Journal of veterinary dentistry, 35*(1), 18–27.
32. McCabe, J. F., & Walls, A. W. (Eds.). (2013). *Applied dental materials.* Wiley.
33. Powers, J. M., & Wataha, J. C. (2015). *Dental Materials-E-Book: Foundations and Applications.* Elsevier Health Sciences.
34. Von Fraunhofer, J. A. (2013). *Dental materials at a glance.* Wiley.
35. Shelton, R. (Ed.). (2016). *Biocompatibility of Dental Biomaterials.* Woodhead Publishing.
36. Tabatabaian, F. (2019). Color aspect of monolithic zirconia restorations: A review of the literature. *Journal of Prosthodontics, 28*(3), 276–287.
37. de la Macorra, J. C., & Pradíes, G. (2002). Conventional and adhesive luting cements. *Clinical Oral Investigations, 6*(4), 198–204.
38. Lad, P. P., Kamath, M., Tarale, K., & Kusugal, P. B. (2014). Practical clinical considerations of luting cements: A review. *Journal of international oral health: JIOH, 6*(1), 116.
39. Wilson, A. D., Paddon, J. M., & Crisp, S. (1979). The hydration of dental cements. *Journal of Dental Research, 58*(3), 1065–1071.
40. Moshaverinia, A., Roohpour, N., Chee, W. W., & Schricker, S. R. (2011). A review of powder modifications in conventional glass-ionomer dental cements. *Journal of Materials Chemistry, 21*(5), 1319–1328.
41. Sidhu, S. K., & Nicholson, J. W. (2016). A review of glass-ionomer cements for clinical dentistry. *Journal of functional biomaterials, 7*(3), 16.
42. Mount, G. J. (2001). *An atlas of glass-ionomer cements: a clinician's guide.* CRC Press.
43. Deb, S., & Chana, S. (2015). Biomaterials in relation to dentistry. In *Biomaterials for Oral and Craniomaxillofacial Applications* (Vol. 17, pp. 1–12). Karger Publishers.
44. Sidhu, S. K. (2011). Glass-ionomer cements restorative materials: a sticky subject? *Australian Dental Journal, 56,* 23–30.
45. Hafshejani, T. M., Zamanian, A., Venugopal, J. R., Rezvani, Z., Sefat, F., Saeb, M. R., ... & Mozafari, M. (2017). Antibacterial glass-ionomer cements restorative materials: A critical review on the current status of extended release formulations. *Journal of Controlled Release, 262,* 317–328.
46. Noumbissi, S., Scarano, A., & Gupta, S. (2019). A literature review study on atomic ions dissolution of titanium and its alloys in implant dentistry. *Materials, 12*(3), 368.
47. Roberts, H. W., Berzins, D. W., Moore, B. K., & Charlton, D. G. (2009). Metal-ceramic alloys in dentistry: A review. *Journal of Prosthodontics: Implant, Esthetic and Reconstructive Dentistry, 18*(2), 188–194.
48. Upadhyay, D., Panchal, M. A., Dubey, R. S., & Srivastava, V. K. (2006). Corrosion of alloys used in dentistry: A review. *Materials Science and Engineering A, 432*(1–2), 1–11.
49. Spencer, P., & Misra, A. (Eds.). (2016). *Material-tissue Interfacial Phenomena: Contributions from Dental and Craniofacial Reconstructions.* Woodhead Publishing.
50. Liu, X., Chen, S., Tsoi, J. K., & Matinlinna, J. P. (2017). Binary titanium alloys as dental implant materials—a review. *Regenerative Biomaterials, 4*(5), 315–323.
51. Gosavi, S., Gosavi, S., & Alla, R. (2013). Titanium: A Miracle metal in dentistry. *Trends in Biomaterials & Artificial Organs, 27*(1).
52. Park, Y. J., Song, Y. H., An, J. H., Song, H. J., & Anusavice, K. J. (2013). Cytocompatibility of pure metals and experimental binary titanium alloys for implant materials. *Journal of Dentistry, 41*(12), 1251–1258.
53. Song, Y. H., Kim, M. K., Park, E. J., Song, H. J., Anusavice, K. J., & Park, Y. J. (2014). Cytotoxicity of alloying elements and experimental titanium alloys by WST-1 and agar overlay tests. *Dental Materials, 30*(9), 977–983.
54. Meffert, R. M., Langer, B., & Fritz, M. E. (1992). Dental implants: a review. *Journal of Periodontology, 63*(11), 859–870.
55. Duraccio, D., Mussano, F., & Faga, M. G. (2015). Biomaterials for dental implants: current and future trends. *Journal of Materials Science, 50*(14), 4779–4812.

56. Özcan, M., & Hämmerle, C. (2012). Titanium as a reconstruction and implant material in dentistry: Advantages and pitfalls. *Materials, 5*(9), 1528–1545.
57. Seo, H. S., Kim, B. H., & Ko, Y. M. (2010). Fabrication of anodized titanium with immobilization of hyaluronic acid to improve biological performance. *Progress in Organic Coatings, 69*(1), 38–44.
58. Reclaru, L., Lüthy, H., Eschler, P. Y., Blatter, A., & Susz, C. (2005). Corrosion behaviour of cobalt–chromium dental alloys doped with precious metals. *Biomaterials, 26*(21), 4358–4365.
59. McCabe, J. F., & Walls, A. W. (Eds.). (2013). *Applied dental materials*. Wiley.
60. Bharti, R., Wadhwani, K. K., Tikku, A. P., & Chandra, A. (2010). Dental amalgam: An update. *Journal of conservative dentistry: JCD, 13*(4), 204.
61. Blackwood, D. J. (2003). Biomaterials: past successes and future problems.
62. Fathi, M., & Mortazavi, V. (2004). A review on dental amalgam corrosion and its consequences.
63. Moncada, G., Fernández, E., Mena, K., Martin, J., Vildósola, P., Junior, O. D. O., ... & Gordan, V. V. (2015). Seal, replacement or monitoring amalgam restorations with occlusal marginal defects? Results of a 10-year clinical trial. *Journal of dentistry, 43*(11), 1371–1378.
64. Corti, C., & Holliday, R. (2009). *Gold: Science and applications*. CRC Press.
65. Knosp, H., Holliday, R. J., & Corti, C. W. (2003). Gold in dentistry: alloys, uses and performance. *Gold Bulletin, 36*(3), 93–102.
66. Kirkup, J. (1993). From flint to stainless steel: Observations on surgical instrument composition. *Annals of the Royal College of Surgeons of England, 75*(5), 365.
67. Vijayalakshmi, R. D., Nagachandran, K. S., Kummi, P., & Jayakumar, P. (2009). A comparative evaluation of metallurgical properties of stainless steel and TMA archwires with timolium and titanium niobium archwires-An in vitro study. *Indian Journal of Dental Research, 20*(4), 448.
68. Corradi, M., Di Schino, A., Borri, A., & Rufini, R. (2018). A review of the use of stainless steel for masonry repair and reinforcement. *Construction and Building Materials, 181*, 335–346.
69. Santander, S. A., & Ossa, C. M. L. (2015). Stainless Steel: Material Facts for the Orthodontic Practitioner. *Revista Nacional de Odontología, 11*(20).
70. Li, H. F., & Zheng, Y. F. (2016). Recent advances in bulk metallic glasses for biomedical applications. *Acta Biomaterialia, 36*, 1–20.
71. Thompson, S. A. (2000). An overview of nickel–titanium alloys used in dentistry. *International Endodontic Journal, 33*(4), 297–310.
72. Xu, X., He, L., Zhu, B., Li, J., & Li, J. (2017). Advances in polymeric materials for dental applications. *Polymer Chemistry, 8*(5), 807–823.
73. Lai, W. F., Oka, K., & Jung, H. S. (2015). Advanced functional polymers for regenerative and therapeutic dentistry. *Oral Diseases, 21*(5), 550–557.
74. Rokaya, D., Srimaneepong, V., Sapkota, J., Qin, J., Siraleartmukul, K., & Siriwongrungson, V. (2018). Polymeric materials and films in dentistry: An overview. *Journal of advanced research, 14*, 25–34.
75. Skirbutis, G., Dzingutė, A., Masiliūnaitė, V., Šulcaitė, G., & Žilinskas, J. (2017). A review of PEEK polymer's properties and its use in prosthodontics. *Stomatologija, 19*(1), 19–23.
76. Rutkuniene, Z., Pervazaite, M., & Skirbutis, G. (2018, September). Modification of Polyetheretherketone Surface by Argon, Oxygen and Nitrogen Plasma for Dentistry Application. In International Conference on Global Research and Education (pp. 160–164). Springer, Cham.
77. Bathala, L., Majeti, V., Rachuri, N., Singh, N., & Gedela, S. (2019). The role of polyether ether Ketone (Peek) in dentistry–A Review. *Journal of medicine and life, 12*(1), 5.
78. Ali, U., Karim, K. J. B. A., & Buang, N. A. (2015). A review of the properties and applications of poly (methyl methacrylate)(PMMA). *Polymer Reviews, 55*(4), 678–705.
79. Wypych, G. (2016). *Handbook of polymers*. Elsevier.
80. Raj, P. A., & Dentino, A. R. (2013). Denture polymers with antimicrobial properties: a review of the development and current status of anionic poly (methyl methacrylate) polymers. *Future medicinal Chemistry, 5*(14), 1635–1645.

81. Paxton, N. C., Allenby, M. C., Lewis, P. M., & Woodruff, M. A. (2019). Biomedical applications of polyethylene. *European Polymer Journal.*
82. Meiers, J. C., Kazemi, R. B., & Donadio, M. (2003). The influence of fiber reinforcement of composites on shear bond strengths to enamel. *The Journal of prosthetic dentistry, 89*(4), 388–393.
83. Vitale, M. C., Caprioglio, C., Martignone, A., Marchesi, U., & Botticelli, A. R. (2004). Combined technique with polyethylene fibers and composite resins in restoration of traumatized anterior teeth. *Dental Traumatology, 20*(3), 172–177.
84. Ganesh, M., & Tandon, S. (2006). Versatility of ribbond in contemporary dental practice. *Trends Biomater Artif Organs, 20*(1), 53–58.
85. Motisuki, C., Santos-Pinto, L., & Giro, E. M. A. (2005). Restoration of severely decayed primary incisors using indirect composite resin restoration technique. *International Journal of Paediatric Dentistry, 15*(4), 282–286.
86. Qualtrough, A. J., & Mannocci, F. (2003). Tooth-colored post systems: A review. *Operative Dentistry, 28*(1), 86–91.
87. Kargul, B., Çaglar, E., & Kabalay, U. (2005). Glass fiber-reinforced composite resin as fixed space maintainers in children: 12-month clinical follow-up. *Journal of Dentistry for Children, 72*(3), 109–112.
88. Freilich, M. A., Meiers, J. C., Duncan, J. P., Eckrote, K. A., & Goldberg, A. J. (2002). Clinical evaluation of fiber-reinforced fixed bridges. *The Journal of the American Dental Association, 133*(11), 1524–1534.
89. Harris, J. M. (1992). Introduction to biotechnical and biomedical applications of poly (ethylene glycol). In Poly (ethylene glycol) Chemistry (pp. 1–14). Springer, Boston, MA.
90. Peng, L., Chang, L., Liu, X., Lin, J., Liu, H., Han, B., et al. (2017). Antibacterial property of a polyethylene glycol-grafted dental material. *ACS Applied Materials & Interfaces, 9*(21), 17688–17692.
91. Zargar, V., Asghari, M., & Dashti, A. (2015). A review on chitin and chitosan polymers: structure, chemistry, solubility, derivatives, and applications. *ChemBioEng Reviews, 2*(3), 204–226.
92. Arnaud, T. M. S., de Barros Neto, B., & Diniz, F. B. (2010). Chitosan effect on dental enamel de-remineralization: an in vitro evaluation. *Journal of Dentistry, 38*(11), 848–852.
93. Lendlein, A., & Kelch, S. (2002). Shape-memory polymers. *Angewandte Chemie International Edition, 41*(12), 2034–2057.
94. Nascimento, R. O. do, & Chirani, N. (2015). Shape-memory polymers for dental applications. In Shape Memory Polymers for Biomedical Applications (pp. 267–280). Woodhead Publishing.
95. Ferracane, J. L. (2011). Resin composite—state of the art. *Dental Materials, 27*(1), 29–38.
96. Curtis, R. V., & Watson, T. F. (Eds.). (2014). *Dental biomaterials: imaging, testing and modelling*. Elsevier.
97. Maas, M. S., Alania, Y., Natale, L. C., Rodrigues, M. C., Watts, D. C., & Braga, R. R. (2017). Trends in restorative composites research: what is in the future?. *Brazilian oral research, 31*.
98. Braga, R. R., Ballester, R. Y., & Ferracane, J. L. (2005). Factors involved in the development of polymerization shrinkage stress in resin-composites: a systematic review. *Dental Materials, 21*(10), 962–970.
99. Leprince, J. G., Palin, W. M., Hadis, M. A., Devaux, J., & Leloup, G. (2013). Progress in dimethacrylate-based dental composite technology and curing efficiency. *Dental Materials, 29*(2), 139–156.
100. García, A. H., Lozano, M. A. M., Vila, J. C., Escribano, A. B., & Galve, P. F. (2006). Composite resins. A review of the materials and clinical indications. *Med Oral Patol Oral Cir Bucal, 11*(2), E215–220.
101. Bayne, S. C. (2005). Dental biomaterials: where are we and where are we going? *Journal of Dental Education, 69*(5), 571–585.
102. Ilie, N., & Hickel, R. (2011). Resin composite restorative materials. *Australian Dental Journal, 56,* 59–66.

103. Moraes, R. R., Goncalves, L. D. S., Lancellotti, A. C., Consani, S., Correr-Sobrinho, L., & Sinhoreti, M. A. (2009). Nanohybrid resin composites: nanofiller loaded materials or traditional microhybrid resins? *Operative dentistry, 34*(5), 551–557.
104. Angerame, D., & De Biasi, M. (2018). Do nanofilled/nanohybrid composites allow for better clinical performance of direct restorations than traditional microhybrid composites?A systematic review. *Operative dentistry, 43*(4), E191–E209.
105. Brunthaler, A., König, F., Lucas, T., Sperr, W., & Schedle, A. (2003). Longevity of direct resin composite restorations in posterior teeth: a review. *Clinical Oral Investigations, 7*(2), 63–70.
106. Opdam, N. J. M., Van De Sande, F. H., Bronkhorst, E., Cenci, M. S., Bottenberg, P., Pallesen, U., ... & Van Dijken, J. W. (2014). Longevity of posterior composite restorations: a systematic review and meta-analysis. *Journal of Dental Research, 93*(10), 943–949.
107. da Veiga, A. M. A., Cunha, A. C., Ferreira, D. M. T. P., da Silva Fidalgo, T. K., Chianca, T. K., Reis, K. R., et al. (2016). Longevity of direct and indirect resin composite restorations in permanent posterior teeth: A systematic review and meta-analysis. *Journal of Dentistry, 54,* 1–12.
108. Tsujimoto, A., Barkmeier, W. W., Fischer, N. G., Nojiri, K., Nagura, Y., Takamizawa, T., ... & Miazaki, M. (2018). Wear of resin composites: Current insights into underlying mechanisms, evaluation methods and influential factors. *Japanese Dental Science Review, 54*(2), 76–87.
109. Zhang, K., Baras, B., Lynch, C. D., Weir, M. D., Melo, M. A. S., Li, Y., ... & Xu, H. H. (2018). Developing a new generation of therapeutic dental polymers to inhibit oral biofilms and protect teeth. *Materials, 11*(9), 1747.
110. Makvandi, P., Jamaledin, R., Jabbari, M., Nikfarjam, N., & Borzacchiello, A. (2018). Antibacterial quaternary ammonium compounds in dental materials: A systematic review. *Dental Materials, 34*(6), 851–867.
111. Scribante, A., Vallittu, P. K., Özcan, M., Lassila, L. V., Gandini, P., & Sfondrini, M. F. (2018). Travel beyond Clinical Uses of Fiber Reinforced Composites (FRCs) in Dentistry: A Review of Past Employments, Present Applications, and Future Perspectives. *BioMed research international, 2018.*
112. Khan, A. S., Azam, M. T., Khan, M., Mian, S. A., & Rehman, I. U. (2015). An update on glass fiber dental restorative composites: a systematic review. *Materials Science and Engineering C, 47,* 26–39.
113. Nicholson, J. W. (2007). Polyacid-modified composite resins ("compomers") and their use in clinical dentistry. *Dental Materials, 23*(5), 615–622.
114. Mahamood, R. M., & Akinlabi, E. T. (2017). *Functionally graded materials* (pp. 16–18). Berlin: Springer.
115. Fujihara, K., Teo, K., Gopal, R., Loh, P. L., Ganesh, V. K., Ramakrishna, S., ... & Chew, C. L. (2004). Fibrous composite materials in dentistry and orthopaedics: review and applications. *Composites Science and Technology, 64*(6), 775–788.
116. Bakar, W. Z. W., Basri, S. N., Jamaludin, S. N. S., & Sajjad, A. (2018). Functionally graded materials: an overview of dental applications. *World Journal of Dentistry, 9*(2), 137–144.
117. Mine, A., De Munck, J., Van Ende, A., Poitevin, A., Matsumoto, M., Yoshida, Y., ... & Van Meerbeek, B. (2017). Limited interaction of a self-adhesive flowable composite with dentin.
118. Poitevin, A., De Munck, J., Van Ende, A., Suyama, Y., Mine, A., Peumans, M., & Van Meerbeek, B. (2013). Bonding effectiveness of self-adhesive composites to dentin and enamel. Dent.
119. Huyang, G., Debertin, A. E., & Sun, J. (2016). Design and development of self-healing dental composites. *Materials and Design, 94,* 295–302.

Chapter 4
Tooth Pastes and Mouthwashes: The Two Commonly Known Dental Preventive Materials

4.1 Introduction

Different kinds of oral care products have been produced since the 1920s. These are used for prevention of teeth from caries and initial mouth problems. Toothpastes, mouthwashes, tooth-whitening products, and so forth [1] are the family of dental preventive materials. Among them, toothpastes and mouthwashes are commonly used among people. As a consequence, studying and having different information about their components, ingredients and so on seems important and this chapter will discuss them in details.

4.2 Toothpaste

Toothpastes date back to around 3,000–5,000 BC when firstly Egyptians developed a compound made by ashes from oxen hooves, myrrh, eggshells, and pumice for cleaning teeth from any debris. Then, Persian people enhanced the toothpaste by addition of burnt shells of snails and oysters alongside with gypsum, herbs, and honey. After around 1,000 years, it was doped by some powder mixture (e.g. crushed bones and oyster shells) for abrasive improvement behavior. Some other additives such as flavor (e.g. ginseng, herbal mints, salt, charcoal, and bark powder) have been added for improved breathing. After the introduction of fluoride in 1914, sodium fluoride was applied in toothpastes for the first time and the first market product of fluoride-rich toothpaste was in 1955 in the USA. After the arrival of sodium lauryl sulfate (SLS) in WWII, it has been used in toothpaste until now.

Toothpastes are likely to be a more complex healthcare product, which contains different additives such as surfactants, active ingredients, flavor compounds, sweeteners, colorings, preservatives, etc. Generally, the ingredients of toothpastes are presented in Table 4.1 [1, 2].

H. Reza Rezaie et al., *A Review on Dental Materials*, Advanced Structured Materials 123, https://doi.org/10.1007/978-3-030-48931-1_4

Table 4.1 Types and portion of compounds in toothpastes [1]

Compounds	Amounts of total	Amount of total/types
Water	20–40%	–
Abrasives	50%	Including aluminum hydroxide, calcium hydrogen phosphates, calcium carbonate, silica, and hydroxyapatite
Fluoride	1450 ppm	In compounds of sodium fluoride, and also Stannous fluoride, sodium monofluorophosphate were included
Detergents	0.5–2%	Usually sodium lauryl sulfate (SLS)
Antibacterial agents	Based on company	Such as triclosan or zinc chloride
Flavorings	Based on company	Spearmint, peppermint, and wintergreen
Remineralizers	Based on company	(In some cases) hydroxyapatite nanoparticles and calcium phosphate
Humectants	Based on company	Glycerol, xylitol, sorbitol, polyethylene glycol, and propylene glycol
Antisensitivity agents	Based on company	(In sensitive cases) strontium chloride and potassium nitrate or arginine
Anticalculus agents	Based on company	Sodium polyphosphate or zinc citrate

During tooth brushing, the formation of the toothpaste slurry has been started with a medium created by saliva and mechanically mixing by the toothbrush. This prepared slurry improves dispersion of active components in the oral cavity, and saliva also changes the slurry pH because of buffering capability. In addition, during this process, due to the reaction between toothpaste excipients the temperature rises. However, other factors like consumer brushing technique and geometry of the used toothbrush are playing an important role in the efficiency of the toothpaste [2].

The color of the tooth is based on two main factors, which are the intrinsic color and extrinsic stains. The former is directly affected by light adsorption and scattering properties of the enamel and dentin. The latter is related to the formation of different materials on the surface of enamel and consequently and account for staining. For instance, colored foods, smoking, and so on lead to extrinsic stains. For addressing these demands, whitening toothpaste could play an important role. The ingredients of this type of toothpaste include solid cleansing abrasive materials, humectant for solubilization of other ingredients, thickening agent to show the rheological characteristics of the formulation, surfactant to produce foam and impart desirable sensorial properties during application, active agents like mentioned fluoride compounds, etc. Table 4.2 illustrates some tooth-whitening materials [3]., *38*, e17–e24.

Moreover, three main toothpaste components are fluoride, SLS, and triclosan, which will be discussed in the following.

Fluoride. The excess of fluoride concentration and consumption (50–100 ppm) especially in drinking water accounts for negative impacts such as neurotoxicity and memory consequences. Likewise, acute systemic toxicity is happening when

Table 4.2 Tooth whitening materials in toothpaste products (Reprinted from Journal of Dentistry, Joiner, A., Whitening toothpastes: a review of the literature, no. e18, Copyright (2010), with permission from Elsevier) [3]

Type of action	Material
Abrasive	Hydrated silica
	Calcium carbonate
	Dicalcium phosphate dihydrate
	Calcium pyrophosphate
	Alumina
	Perlite
	Sodium bicarbonate
Chemical	Hydrogen peroxide
	Calcium peroxide
	Sodium citrate
	Sodium pyrophosphate
	Sodium tripolyphosphate
	Sodium hexametaphosphate
	Papain
Optical	Blue covarine

fluoride-rich toothpaste is taken [1]. Several fluoride compounds are used based on the country. For instance, in the EU fluoride ingredients are ingesting by different market products especially as a cosmetic material. In the EU, various types of fluoride mixtures with a maximum dose of 1,500 ppm are allowed. However, in the USA, only a few types and a low amount of fluoride dose are allowed. Table 4.3 presents information about types of permitted fluoride compounds in the USA and EU [2].

The maximum permissible fluoride concentration in toothpaste, however, is related to the age of the consumer and is different for children and adults. For children under the age of 6, the fluoride concentration in toothpaste is between 250 and 1,500 ppm in the EU. But in the USA, the maximum of that is 1,150 ppm for 2 years and older [2].

Sodium Lauryl Sulfate (SLS). For improvement in cleansing actions, sodium lauryl sulfate (SLS) is added which is the most common surfactant with chemical linear formula of $C_{12}H_{25}NaO_4S$. SLS is sodium salt of lauryl alcohol (1-dodecanol) and known as sulfuric acid monododecyl ester sodium salt. The surfactant helps the dispersion of toothpaste in the oral cavity and in the micellization of hydrophobic compounds (e.g. flavor ingredients and antiplaque-antigingivitis actives). Moreover, SLS has different advantages such as inhibition of microorganism growth which is related to the penetration into porous cell wall in parallel, interaction with cell membrane, lipids, and proteins. According to the researches, SLS could penetrate deeply into biofilms. However, SLS may show negative impact on oral health by recurrent aphthous ulcers (RAU) like elimination of the protective mucin surface layers and consequently, reduction of oral mucosa resistance. For this, it was mentioned that patients who are suffering from dry mouth or sensitive mucosa limits

Table 4.3 Permissible compounds and does of fluoride for application in cosmetic and/or toothpaste according to the EU and USA terms and conditions [2]

Compounds	Linear formula	Comments
EU (commonly used in cosmetic products)		
Amine fluoride 3-(N-hexadecyl-N-2-hydroxyethyl-ammonio) propylbis(2-hydroxy-ethyl) ammonium dihydrofluoride	$C_{27}H_{59}N_2O_3F_2$	–
Aluminum fluoride	AlF_3	–
Ammonium monofluorophosphate	$(NH_4)_2PO_3F$	–
Ammonium fluoride	NH_4F	–
Ammonium fluorosilicate	$(NH_4)_2SiF_6$	–
Calcium fluoride	CaF_2	–
Calcium monofluorophosphate	$CaPO_3F$	–
Hexadecyl ammonium fluoride	$C_{16}H_{35}NHF$	–
Magnesium fluoride	MgF_2	–
Magnesium fluorosilicate	$MgSiF_6$	–
N,N′,N′-tris(polyoxyethylene)-N hexadecyl-propylenediamine Dihydrofluoride	n/a	–
Nicomethanol hydrofluoride	C_6H_8FNO	–
Octadecenyl-ammonium fluoride	n/a	–
Potassium fluoride	K/F	–
Potassium fluorosilicate	K_2SiF_6	–
Potassium monofluorophosphate	K_2PO_3F	–
Sodium fluoride	NaF	–
Sodium fluorosilicate	Na_2SiF_6	–
Sodium monofluorophosphate	Na_2PO_3F	–
Tin difluoride	SnF_2	–
USA (commonly used in toothpaste)		
Sodium fluoride	NaF	Theoretical fluoride allowable concentration: 850–1,150 ppm Minimum access fluoride concentration: 650 ppm
Sodium monofluorophosphate	Na_2PO_3F	Theoretical fluoride allowable concentration: 850–1150 or 1500 ppm Minimum access fluoride concentration: 800 or 1,275 ppm
Stannous fluoride	SnF_2	Theoretical fluoride allowable concentration: 850–1,150 ppm Minimum access fluoride concentration: 700 or 290 for products containing calcium pyrophosphate

the usage of SLS-rich toothpaste and any oral care products. Additionally, SLS-containing products could lead to a hypersensitivity to metal ions with escalating the mucosal permeability to sensitizing ingredients. Nowadays, different oral care products and toothpastes are introduced without having any components of SLS [4, 5].

Triclosan. It is antimicrobial agent with the chemical name of 2,4,4′-trichloro-2′-hydroxy-diphenyl ether and also common as Irgasan DP300, FAT80′023, CH 3565, GP41-353, and Irgacare MP (the pharmaceutical grade of triclosan that is >99% pure). It has been used since 1968 which was firstly applied in dentistry around 1980 in the EU. Triclosan was firstly approved by the US Food and Drug Administration (FDA) in 1990. Triclosan has a wide variety of applications in care products such as toothpaste, antibacterial soaps, deodorant soaps, cosmetics products, and so on. Plus, triclosan has been observed in urine plasma and breast milk in humans [6]. It was shown that triclosan could induce apoptosis in human gingival fibroblasts, but it can drop side effects of SLS on oral mucosa. Triclosan has negative impact on the function of thyroid gland if applied for a long time. However, in general, triclosan is considered as a low allergenic potential and 0.3% triclosan in toothpaste is allowable and side effect free on thyroid. In some cases, it was studied that triclosan/copolymer usage in toothpaste has not any serious negative consequences on plaque, gingivitis, and caries controlling [1].

4.3 Mouthwashes

Mouthwashes or mouth rinses are a type of oral home care products employed for the health and care of the teeth and the mouth [1]. Mouthwashes are generally flavored solutions and can be also used for freshening the breath; however, the use of these products does not replace brushing or flossing [7]. Mouthwashes have been produced in various types based on the main active components in order to provide specific indications. In this manner, fluoride mouthwashes are one of the mostly used ones, which contain approximately 0.05% sodium fluoride (NaF) and are suitable for daily use. These kinds of mouthwashes protect teeth from formation or development of caries especially for patients who require excess amounts of fluoride due to the great risk of caries formation in teeth. Antiplaque mouthwashes, on the other hand present some undesirable effects when used regularly, thus they should be used in particular conditions in order to decrease the plaque and biofilm accumulation on the surface of the teeth. Chlorhexidine, essential oils, cetylpyridinium chloride (CPC), and triclosan are antiseptic agents added to the composition of antiplaque mouthwashes. Desensitizing mouthwashes are the last type of mouthwashes containing 3–5 wt% potassium nitrate or 0.8 wt% arginine to minimize dentine hypersensitivity of patients suffering from that [1].

Chlorhexidine. Chlorhexidine is an ideal broad-spectrum bisbiguanide-based cationic antiseptic introduced in the 1950s and is widely used nowadays [8]. It is applied in the forms of digluconate, hydrochloride, and acetate. Chlorhexidine protects oral health against organisms such as Gram-positive bacteria, Gram-negative bacteria, and fungi as well as killing bacteria causing infection [7]. This antimicrobial agent is commercially available as mouthwash/rinse (0.06, 0.1, 0.12, and 0.2%), toothpaste (1%), gel (1%), spray (0.2%), root canal irrigant (2%), varnishes, and periodontal chips for short-term use [1, 8]. Chlorhexidine is also used in several clinical applications for caring oral hygiene after periodontal treatment and orthognathic surgery as well as in patients with blood problems after chemotherapy, radiotherapy, and bone marrow transplants. It is also applied for treating oral malodour [1]. Besides oral applications chlorhexidine is also used for preventing disease transmission as well as preventing infections of minor cuts, grazes, burns, scalds, blisters, insect bites, and min or infections of the mouth or throat [7]. The capacity of chlorhexidine in providing antiplaque properties and preventing development of gingivitis has been discussed in many in vitro studies [1]. A study carried out on adults in the USA has shown that about 50% of population face gingivitis problems of about 3–4 teeth, which can be treated to some extent using chlorhexidine gluconate oral rinse [7]. Many studies have shown significant decrease of plaque and biofilm formation after four to six months and six months in patients who use mouth rinses twice a day. However, it was demonstrated in some cases that after four weeks of continuous utilization of chlorhexidine rinse brings about teeth staining. In addition, studies have shown that there is no significant difference between different concentrations of chlorhexidine rinses in their functional efficiency [8]. The functional mechanism of cationic chlorhexidine includes the formation of bonds with an anionic component present in the oral environment such as salivary glycoproteins, free carboxyl, sulfate, and phosphate groups, which limits adhesion of compounds such as proteins to the enamel. Thus, providing antimicrobial properties and minimizing biofilm and plaque formation over the teeth surface [1]. The functional mechanism of chlorhexidine involves bacteriostatic (inhibition of bacterial growth) and bactericidal (killing bacteria) actions associated with its concentration [7]. The major disadvantages of chlorhexidine are teeth, mucosa, and restorations discoloration as well as taste disturbance, parotid swelling, oral mucosal erosion, and increased formation of supragingival calculus. Allergic reactions have been mentioned in some cases with the use of chlorhexidine gluconate gel and mouthwash. As a matter of fact, due to the hypersensitivity reactions and probable teratogenic potential of chlorhexidine gluconate, its use is restricted during pregnancy [1].

Essential Oils. Essential oils-containing mouthwashes found to have similar efficiency to chlorhexidine. Thymol (0.064%), menthol (0.042%), eucalyptol (0.092%), methyl salicylate (0.06%), and tea-tree leaf oil are the essential oils frequently used in mouthwashes. Taste alteration, epithelial sloughing, and erythema are the only inconvenient effect of these mouthwashes. Essential oil mouthwashes barely present allergic reactions and the clinical studies performed on the efficacy and safety of them exhibit outstanding benefits without any important reactions with oral mucosa [1].

Alcohol. Ethanol is an antiseptic and preservative liquid used in many mouthwashes acting as a solvent of active agents. As alcohol-containing mouthwashes normally have high concentrations of ethanol and are in direct contact with the oral mucosa within long periods, their side effects should be thoroughly considered. The probable side effects of alcohol-containing mouthwashes are burning sensation, oral mucosa drying, and softening of composite fillers. In addition, there is a great source of issue regarding the carcinogenic potential of mouthwashes, which contain alcohol. There has been some disagreement concerning whether there is any relation between alcohol-containing mouthwashes and oral cancer risk. In fact, some studies demonstrate the enhanced risk of oral cancer with the use of ethanol-containing mouthwashes. However, the American Dental Association in March 2009 declared that these mouthwashes are safe and effective in preventing plaque formation on the teeth for individuals with healthy lifestyle and those who do not have an intolerance for alcohol and do not smoke. To conclude, until now there is not any decisive evidence proving or disproving carcinogenic effect of ethanol-containing mouthwashes. As a matter of fact, it is wise to minimize the long-term use of high alcohol-containing products [1, 9].

Benzydamine. Benzydamine hydrochloride is added to some mouthwashes due to its anti-inflammatory, and antimicrobial properties, which can minimize pain and inflammation related to oral ulceration, stomatitis, aphthous ulcers, radiotherapy-induced mucositis, and oral surgery. Occasional oral mucosa paresthesia and photoallergic contact dermatitis of lips and chin are probable drawbacks of the use of benzydamine mouthwashes [1].

Povidone Iodine. Povidone iodine (polyvinylpyrrolidone-iodine) also called Betadine (1%) mouthwash has great antibacterial property used for treating severe oral mucosal infections and decreasing bacterial concentration in the oral cavity. These mouthwashes may cause type-I hypersensitivity allergic reaction, local mucosal inflammation, thyroid function disability, metabolic acidosis, and disabled renal function [1].

Cetylpyridinium Chloride. Cetylpyridinium chloride (CPC) is a cationic quaternary ammonium presenting antibacterial and antiplaque properties, which is used in some mouthwashes with low weight percentages (0.07 wt%). The adverse effects of these mouthwashes are discoloration of teeth surface, altered taste, and burning sensation [1].

Sodium Bicarbonate. Sodium bicarbonate ($NaHCO_3$) limits bacteria growth and biofilm formation on the teeth surface by increasing the pH in the oral cavity. It also cleanses oral environment by debris loosening as well as mucus dissolving. A clinical study on sensitive oncology patients has demonstrated that 1% sodium bicarbonate mouthwash provides inferior irritation compared to chlorhexidine mouthwashes. These mouthwashes have low cost and are available and permissible for daily use [1].

Delmopinol Hydrochloride. Delmopinol hydrochloride is a cationic agent with activated surface used in some mouthwashes with the ability of disrupting plaque structure and limiting bacterial accumulation on the surface of teeth. The adverse effects of delmopinol are teeth staining and temporary lingual paresthesia [1, 10].

Hydrogen Peroxide. Hydrogen peroxide is an oxidizing agent used in the composition of some mouthwashes (1.5 wt%) for short-term uses for eliminating anaerobic bacteria presented in necrotizing ulcerous gingivitis and periodontitis. Bubble production of hydrogen peroxide also has a mechanical cleansing effect. Hydrogen peroxide presents in vitro genotoxic potential with directly contacted living organisms. In contrary, in vivo genotoxicity has not been seen due to the presence of metabolizing enzymes, which limit genotoxic effects. Despite the fact that hydrogen peroxide may cause lingual papilla growth; however, low concentrations are rather safe to use [1, 11].

References

1. Moharamzadeh, K. (2017). Biocompatibility of oral care products. In *Biocompatibility of dental biomaterials* (pp. 113–129). Woodhead Publishing.
2. Lippert, F. (2013). An introduction to toothpaste-its purpose, history and ingredients. In *Toothpastes* (Vol. 23, pp. 1–14). Karger Publishers.
3. Joiner, A. (2010). Whitening toothpastes: A review of the literature. *Journal of Dentistry, 38,* e17–e24.
4. Sälzer, S., Rosema, N. A. M., Martin, E. C. J., Slot, D. E., Timmer, C. J., Dörfer, C. E., et al. (2016). The effectiveness of dentifrices without and with sodium lauryl sulfate on plaque, gingivitis and gingival abrasion—A randomized clinical trial. *Clinical Oral Investigations, 20*(3), 443–450.
5. Marsh, P. D. (2010). Controlling the oral biofilm with antimicrobials. *Journal of Dentistry, 38,* S11–S15.
6. Rodricks, J. V., Swenberg, J. A., Borzelleca, J. F., Maronpot, R. R., & Shipp, A. M. (2010). Triclosan: A critical review of the experimental data and development of margins of safety for consumer products. *Critical Reviews in Toxicology, 40*(5), 422–484.
7. Kumar, S. B. (2017). Chlorhexidine mouthwash-a review. *Journal of Pharmaceutical Sciences and Research, 9*(9), 1450.
8. Richards, D. (2017). Chlorhexidine mouthwash plaque levels and gingival health. *Evidence-based Dentistry, 18*(2), 37.
9. Reidy, J., McHugh, E., & Stassen, L. F. A. (2011). A review of the relationship between alcohol and oral cancer. *The Surgeon, 9*(5), 278–283.
10. Addy, M., Moran, J., & Newcombe, R. G. (2007). Meta-analyses of studies of 0.2% delmopinol mouth rinse as an adjunct to gingival health and plaque control measures. *Journal of clinical periodontology, 34*(1), 58–65.
11. Walsh, L. J. (2000). Safety issues relating to the use of hydrogen peroxide in dentistry. *Australian Dental Journal, 45*(4), 257–269.

Chapter 5
Nanoparticles (NPs) in Dentistry

5.1 Introduction

The term nanotechnology, referring to technologies operating at very small scales, was first introduced by Richard Feynman. The structural size of nanomaterials is between 1 and 100 nm which is attributed to special mechanical and chemical properties that are more powerful than those used on a bulk scale [1]. This field started to develop in the 1890s by the arrival of cluster science and the evolution of scanning tunneling microscope (STM). Additionally, the publicizing of nanotechnology was done by K Eric Drexler around 1986 in his published book Engines of Creation [2]. Even, the first researcher who considered the possible usage of nanotechnology to medicine was Richard Feynman for curing heart disease: "a friend of mine suggests an interesting possibility for relatively small machines. He says that, although it is a very wild idea, it would be interesting in surgery if you could swallow the surgeon. You put the machine surgeon inside the blood vessel and it goes into the heart and looks around". In Table 5.1, the developments of nanotechnology in the last eight decades are summarized based on countries and events. In Table 5.2, nanomaterials are classified based on their dimensions in four main groups by Siegel [3]. This development in dentistry which so-called "nanodentistry" will make possible the maintenance of near-perfect oral cavity health and is not farfetched imagination [4]. As a result, different studies have been conducted on the application of nanotechnology in dentistry. For instance, the effect of glass nanoparticles (NPs) on the filling composites, using of nano-carbonate apatite on the re-staining after the bleaching process, impacts of nano-oxides on the color stability of pigmented maxillofacial silicone elastomer, etc. [5].

Nowadays conventional dental materials become outdated due to the side effects and drawbacks. For instance, although amalgam is famous because of good mechanical properties which are used for sealing crowns and bridges, it contains mercury that has a negative side effect. Likewise, other conventional materials like glass ionomers

H. Reza Rezaie et al., *A Review on Dental Materials*, Advanced Structured Materials 123, https://doi.org/10.1007/978-3-030-48931-1_5

Table 5.1 Historical evolution in nanotechnology (Reprinted from Springer, Cham, Chaughule, R. S. (Ed.), Dental Applications of Nanotechnology, no. 84, Copyright (2018), with permission from Springer) [3]

Number	Year	Events	Region/people
1	1931	Transmission electron microscope (TEM)	M. Knoll and E. Ruska (Technical University of Berlin, Germany)
2	1937	Scanning electron microscope (SEM)	M. vonArdenne (Forschungslaboratorium für Elektronenphysik, Germany)
3	1954	Targeted drug delivery concept by miniature particles—magic bullets	Paul Ehrlich
4	1959	Feynman's Lecture on "There's Plenty of Room at the Bottom"	R.P. Feynman (California Institute of Technology, Pasadena, CA, USA)
5	1960s	Investigation of polyacrylic beads for oral administration	Peter Paul Speiser, Zurich
6	Late 1960s	First nanoparticles for drug delivery purposes and for vaccines	Paul Speiser's research group
7	1969	"Micelle polymerization"—First work employed toward processing sustained release nanoparticles for vaccination purpose	Gerd Birrenbach, Germany
8	1974	Concept of nanotechnology proposed	N. Taniguchi (Tokyo University of Science, Japan)
9	1977	Lysosomotropic effects of the nanoparticles discovered	Patrick Couvreur, Zurich
10	1978	First review article about nanoparticles	Kreuter, Germany
11	1979	First rapidly biodegradable acrylic nanoparticles	Patrick Couvreur, Zurich
12	1981	First poly (lactic acid) nanoparticles for drug delivery developed	Robert Gurny, Switzerland
13	1986	Three-dimensional space manipulation of atoms Demonstrate	S. Chu (Bell Lab., USA)
14	1984, 1986	Nanoparticles can improve the oral bioavailability of drugs	Maincent et al., France
15	1986	Nanocapsules	Al Khouri Fallouh et al.
16	1993	Concept of nanomedicine	Robert A. Freitas Jr.
17	1995	Nanoimprinting	S.Y. Chou, USA
18	1996	Nanosheets	T. Sasaki (National Institute for Research in Inorganic Materials, Japan)

(continued)

Table 5.1 (continued)

Number	Year	Events	Region/people
19	2000	Term "nanodentistry" introduced	Robert Freitas, USA
20	2000	National Nanotechnology Initiative (NNI)	USA
21	2003	Twenty-first Century Nanotechnology Research and Development Act	USA
22	2005	Nanosciences and Nanotechnologies: An action plan	Europe

Table 5.2 Nanomaterials classification based on dimensions (Reprinted from Springer, Cham, Chaughule, R. S. (Ed.), Dental Applications of Nanotechnology, no. 179, Copyright (2018), with permission from Springer) [3]

Dimension	Characteristics	Examples
Zero	Clusters/powders	Atomic clusters, filaments, and cluster assemblies
One	Multilayers	Nano-thin film
Two	Ultrafine grained over-layers or buries layers	Nanotubes, nanofibers, nanowires
Three	Nanophase materials consisting of equiaxed nanometer-sized grains	Nanoparticles, nanopowders, dendrimers, fullerenes, quantum dots, nanostructure, nanopores

for temporary filling, gold for prosthesis, ceramics for crowns and bridges application, steel for frameworks, etc. show damaging impacts, are presented in Table 5.3 [6].

5.2 Family of Nanoparticles in Dentistry

The nanometric size range with different morphologies includes spherical, cubical, or needle-shaped particles that offer not only excellent mechanical properties but also antibacterial activities due to the high surface to volume ratio. In Table 5.4 widely used nanoparticles by their applications are presented which are silver, gold, iron oxide, copper oxide, hydroxyapatite, titanium dioxide, calcium fluoride, calcium phosphate, silica, chitosan, zirconia, zinc oxide, PLGA, carbon nanotube (CNT), and graphene. These NPs are discussed in the next following subsections.

Silver. Silver nanoparticles or AgNPs have been attracted by scientists because of unique physical and biochemical properties rather than related macro- and micro-components. These materials can battle against 650 different infection-causing organisms. AgNPs can diffuse to the cell membranes of biofilm organisms and causing cell death where other antibacterial agents have not this capability. However, different

Table 5.3 Pros and cons of conventional dental materials (Reprinted from Journal of Oral Biology and Craniofacial Research, Priyadarsini et al., Nanoparticles used in dentistry: A review, no.2, Copyright (2018), with permission from Elsevier) [6]

Conventional materials	Benefits	Drawbacks
Amalgam	Durable, provide great resistance to surface corrosion, easy to manipulate, needs less time to get placed than other materials; prevention of bacterial leakage, lasts long when placed under controlled conditions, money saving	Disruption of the tooth tissue, reduced esthetic qualities, local allergic responses may occur, mercury which is component may have toxic effects
Nickel or cobalt chrome alloys	Good resistance to further decay, long-term effectivity, shows no fracture issues in stress conditions, does not decay in the mouth, and needs the removal of the minimum tooth, resistance against leakage	Dark silver metal-colored may irritate sensitivity tooth as it conducts heat and cold, costly
Glass ionomer	Provide esthetic beauty, prevents decay by fluoride release, needs the removal of the minimum tooth, low tooth sensitivity	Usage is limited, the material becomes rough as time goes and can increase the accumulation of plaque resulting in periodontal disease
Gold alloy	Good resistance to decay, durable, shows no fracture in stress, resists leakage because of good fitting, needs the removal of the minimum tooth	It is colored yellow, may result in tooth sensitivity, high cost
Ceramics	Provide excellent resistance to further decay, resistant to surface wear, resistant to leakage, very minimal or no tooth sensitivity	The weak material, which can break under biting forces, not suggested for molar teeth, costly material
Ceramics fused to metals	Resistance to decay, very durable, no sensitivity of tooth, resists leakage	More tooth needs to be removed than ceramics, higher cost
Resin ionomer	Very good esthetics, act against decay by releasing fluoride, needs the removal of the minimum tooth, well usable for non-biting surfaces, can be used for the primary restoration of teeth, better than glass ionomers, provide resistance against leakage, low rate of dental sensitivity	Cost is similar to the composite resin, limited use wears faster than composite and amalgam

(continued)

Table 5.3 (continued)

Conventional materials	Benefits	Drawbacks
Composite resin	Strong and durable, tooth-colored, break free, preservation of maximum tooth, the low hazard of leakage, do not undergo decay and provide good resistance to the biting forces	Moderate tooth sensitivity occurs, more costly, shrinks when it hardened which could lead to temperature sensitivity, may leak over time

mechanisms have been introduced for AgNPs which are antibacterial, antiviral, and antifungal activities. In terms of antibacterial mechanism, silver ions are attributed to interruption of ATP molecules and preventing DNA replication, reactive oxygen formation species, or dell membrane damage. So, AgNPs create holes in the cell wall of gram-negative organisms and causing cell death by increasing permeabilities. Figure 5.1 summarizes the related process of antibacterial actions of AgNPs. In most cases, the antibacterial activities of AgNPs happen due to the Ag^+ flow out and can be more productive when the AgNPs size is <10 nm.

About antiviral activities, AgNPs inhibit viruses binding and replication and this process should gain attention more for complications. Antifungal actions for AgNPs illustrated battle against 44 strains of different fungal species. For instance, AgNPs can destruct the cell membrane of Candida albicans and consequently ban cell growth. Different applications of AgNPs in dentistry have been introduced including acrylic resins, composite resins and adhesive, endodontics, periodontal materials, porcelain restorations, titanium implants, orthodontics materials, and cancer treatments. In Table 5.5 the influence of AgNPs cooperated by various dental biomaterials is discussed [10].

Gold. Nowadays, gold is used in medical devices including pacemakers and gold-plated stems, controlling of disease, middle ear gold implants, and gold alloys in dental restoration [11]. Gold nanoparticles or GNPs properties are the wine-red solutions at which their properties will be affected by GNPs interactions. The range of GNPs size changing from 1 nm to 8 μm, with different shapes (Fig. 5.2) like a spherical ring, sub-octahedral, icosahedral tetrahedral, decahedral, octahedral, and nanorods [12] which are capable for easily synthesis through a one-step environmentally friendly chemistry method that is applicable for its biocompatibility and non-toxic actions. So, GNPs will be an excellent choice for biomedical incorporation. Chemical methods for production are using a wide variety of reducing agents (e.g. sodium borohydride) at which the most related drawbacks are the generation of a large number of hazardous by-products. As a consequence, in the last decades, green methods have gained more attention which are biocompatible and eco-friendly. Different plants like Nepenthes khasiana, Pterocarpus marsupium, and Salacia Chinensis L (SC) with high antioxidant properties are used for green synthesis of GNPs. GNPs have been applied in the form of various coatings agents which play a role in antibacterial and antifungal actions. Additionally, optimum concentration and

Table 5.4 Important nanoparticles (NPs) used in dentistry based on properties, applications and synthesis methods [1, 6–9]

Nanoparticles	Preparation routs	Properties alongside of applications
Silver	Combination of Ag ethyl hexanoate and ethyl methacrylate, Catalytic chemical vapor deposition (CVD) with the addition of distilled water	Reduction of biofilm formation, enhanced mechanical properties of nanocomposites and at lower concentrations, AgNPs solution had a bactericidal effect similar to 5.25% NaOCl; thus, can be used as intracanal irrigates. It could be used in dental implants and prosthetics. In terms of side effects, AgNPs induce toxicity. Chronic exposure to silver can have argyria. Toxic effects of AgNPs are due to ROS production. Both silver ions and AgNPs contribute to the toxicity. Ag NPs involved in oxidative stress production and genotoxicity, lysosomal AcP activation, actin disruption, hemocytes phagocytosis stimulation, and inhibition of Na-K-ATPase
Gold	Solvent evaporation method, Reduction of chloroauric acid with trisodium citrate, Photothermal reduction procedure	Increasing concentration of AuNP improved mechanical properties of dentin adhesive did not show any cytotoxicity but high antibacterial activity, enhancing bone formation and maintenance and can be used as contrast enhancers for imaging techniques
Iron oxide	Thermal decomposition method, Iron chloride combined with sodium hydroxide	Combinational strategies, such as polymer brush coating to biomaterial surface and iron oxide NPs, significantly decreased biomaterial associated infections, SPIONs alone or in combination prevent biofilm formation, Size-dependent effects on cellular toxicity, and inflammatory response

(continued)

Table 5.4 (continued)

Nanoparticles	Preparation routs	Properties alongside of applications
Copper oxide	The solvothermal method, Tesima^{TM} plasma process	Addition of CuONPs in concentrations up to 1% in a two-step etch-and-rinse adhesive system was feasible, and could provide antimicrobial properties and preserve bonding to dentin after 1 year, without reducing mechanical properties of adhesive formulations, decreased bacterial load and inhibited the growth of biofilm organisms, efficient against bacteria associated with hospital-acquired infections
Hydroxyapatite	Dry & wet methods, high-temperature processes, preparation from biogenic sources and combination procedures	Improve bond strength, HAP NPs paste proved high efficiency for the reduction of sensitivity after tooth whitening, similar composition with teeth and bone, biocompatible, absorbed to the enamel of the teeth, reform periodontal shortcomings, making a film of artificial enamel around the tooth for protection and easily integrated into the dental tubules
Titanium dioxide	Sol–gel technique	Improved antibacterial and physical properties of GI restorative material, enhanced mechanical properties of the resin, long-term effect on dental implants, application in dental surfaces with less bacterial adhesion consequence and hardness improvement
Calcium fluoride	Spray drying process	Anti-caries and desensitizing agent, CaF_2 composites with CaF_2 NPs have high strength and sustain release of fluoride ions
Calcium phosphate	Spray drying process	Novel rechargeable CaP composites release Ca and P ions, CaP recharge method improved mechanical and antibacterial properties

(continued)

Table 5.4 (continued)

Nanoparticles	Preparation routs	Properties alongside of applications
Silica	Sol–gel methods with silanization modification, Hydrophobic SiO_2 coated with a silane coupling agent	Best mechanical properties achieve at low nanofiller concentrations where the dispersion of NPs is more homogeneous, SiO_2 NPs enhance mechanical properties of elastomer, increase fracture toughness of restorations, biocompatible, low toxic side effects, low density and an impressive adsorption ability and cost-effective and polishing agent for softening the surface of teeth
Chitosan	Water in oil method, emulsion-droplet coalescence technique, and spray drying process	Improve cell response observed in terms of attachment and proliferation because of the addition of chitosan nanofibers, enhance adhesion properties, low cytotoxicity and continuous release of chlorhexidine observed
Zirconia	Laser vaporization	NPs in the adhesive layer or in primer increase tensile strength and promoted mineralization, the addition of ZrNPs enhanced flexural strength but lowered impact strength with increased concentrations of ZrNPs, the similarity in mechanical properties and color with the tooth, low cytotoxicity, sensible biocompatibility, high fracture resistance
Zinc oxide	Chemical synthesis	Prevent bacterial colonizers responsible for oral biofilm formation
PLGA	Ion pairing method	Exhibit antibacterial action against organism responsible for root canal infections, incorporation of minocycline with PLGA NPs showed antibacterial effects specifically against Aggregatibacter Actinomycetemcomitans
Carbon Nanotube (CNT)	Arc discharge, laser ablation, and chemical vapor deposition (CVD)	Teeth filling and coatings with large surface area, bring active agents to live cells, adheres easily to the tooth surfaces and to the surfaces of dentin and cementum

(continued)

Table 5.4 (continued)

Nanoparticles	Preparation routs	Properties alongside of applications
Graphene	Chemical reduction of graphite oxide with reducing agents like Cocos nucifera L. (coconut water), exfoliation of graphite oxide under strongly alkaline conditions, low-temperature synthesis using sodium ammoniate and chemical vapor deposition (CVD)	Teeth coating, suitable for implantation, biofilm reduction, cost-effectiveness, fracture resistance, low density from a uniform crystal lattice, and treat bacterial biofilm

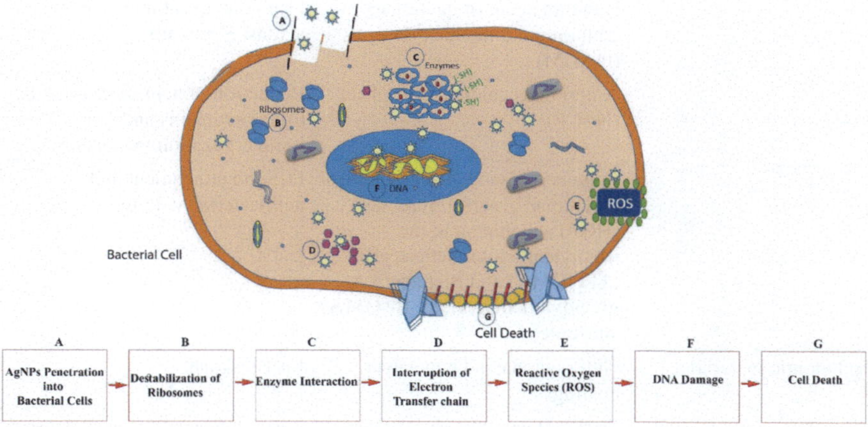

Fig. 5.1 Steps of antibacterial activities when coming across a bacterial cell (Reprinted from Materials Science and Engineering: C, Bapat et al., An overview of application of silver nanoparticles for biomaterials in dentistry, no. 883, Copyright (2018), with permission from Elsevier) [10]

particle size of GNPs illustrated five and ten times for tensile and flexural strength, respectively, in terms of mechanical behaviors. However, the film composition of AuNPs with chitosan proved high antibacterial actions against antibiotic-resistant strains of Pseudomonas aeruginosa and Staphylococcus aureus. GNPs are also applicable for the promotion of bone regeneration by coating on a titanium implant and are useful for the preservation of nascent bone formation [1].

Iron Oxide. Iron compounds (FeO_x) with high-surface areas play an important role in different fields like contrast enhancement in magnetic resonance imaging (MRI), repairing of tissue, hyperthermia treatment, and antibacterial activities [1]. Magnetite, maghemite, and superparamagnetic nanoparticles are mostly common in biomedical applications [6]. Superparamagnetic iron oxide nanoparticles (SPIONs) are capable for powering osteoblast functions and efficiently fight against biofilms formed over the surfaces of the dental implants which are difficult to get rid of,

Table 5.5 Incorporation of Ag nanoparticles (NPs) with different materials based on their applications [10]

Dental applications	Incorporation	Presence of AgNPs
Acrylic resins for dentures	Poly (methyl methacrylate) (PMMA) and bisphenol (glycidyl methacrylate resin)	Decrease C. Albicans adherence, antibacterial action against S. aureus, Pseudomonas aeruginosa and S. Mutans, improvement in the mechanical properties
Composite resins	Calcium silicate cements like Mineral trioxide aggregate (MTA) and Portland Cement (PC)/Zirconium oxide	High resistance for compression due to the decreased porosity, higher oral antibacterial resistance (e.g. S. mutans)
	Bonding agent of quaternary ammonium dimethacrylate (QADM)	Depiction of inhibitory role against S. mutans
	Amorphous calcium phosphate (NACP)	Reduction of acid production by microorganisms and remineralization increasing
	Influence of nanosilver Coatings on Human gingival fibroblast (HGF) attachment to the triethylenglicoldimethacrylate (TEGDMA) and Bisphenol A glycidyl methacrylate (BisGMA) substrates	Definite attachment and antibacterial activities
Endodontic materials	Gutta-percha and nanosilver coatings	Less leakage
	Sodium hypochlorite (NaOCl)	Less concentration needed and better microorganism, elimination like E. faecalis, S. aureus
	Mineral trioxide aggregate (MTA)	High antibacterial and antifungal activities
Periodontology	Carboxymethyl cellulose	More effective at prevention of gram-negative organism* growth
Porcelain for Crowns and Bridge	Porcelain	Shield against subcritical crack growth and high fracture resistance
Titanium dental implants	Silver plasma immersion (Ag-PIII) and pure titanium	High inhibition capabilities against bacterial (S. aureus and E. coli)
	Hydrogel titanate ($H_2Ti_3O_7$) nanotube-layered titanium foils	High bacteriostatic activity pf 99.99% and long-term Ag + release potential

(continued)

Table 5.5 (continued)

Dental applications	Incorporation	Presence of AgNPs
	Chitlac-nAg coated on BisGMA/TEGDMA specimens	Moderate antibacterial effects (against S. aureus and P. aeruginosa), acceptable biocompatibility, no significant adverse bone reaction
	Sandblast large grit (SLA) and acid etch treated titanium	Inhibition growth of F. nucleatum and S. aureus
	AgNPs-containing multilayer coating on phase-transited lysozyme-primed (PTL) Titanium surfaces	Promoted early osseointegration of the implant
Orthodontic cement and adhesives	Composite adhesive containing silica nanofibers	Reduce the adhesion of streptococci to the orthodontic adhesive
	Hydroxyapatite (HA) amalgamated in adhesive	Excellent antibacterial actions, increasing shear bond strength (SBS)
Anticancer treatment	Chitosan-alginate	Increase in apoptotic cell levels

*Gram-negative organisms are like E. coli and P. aeruginosa bacterial

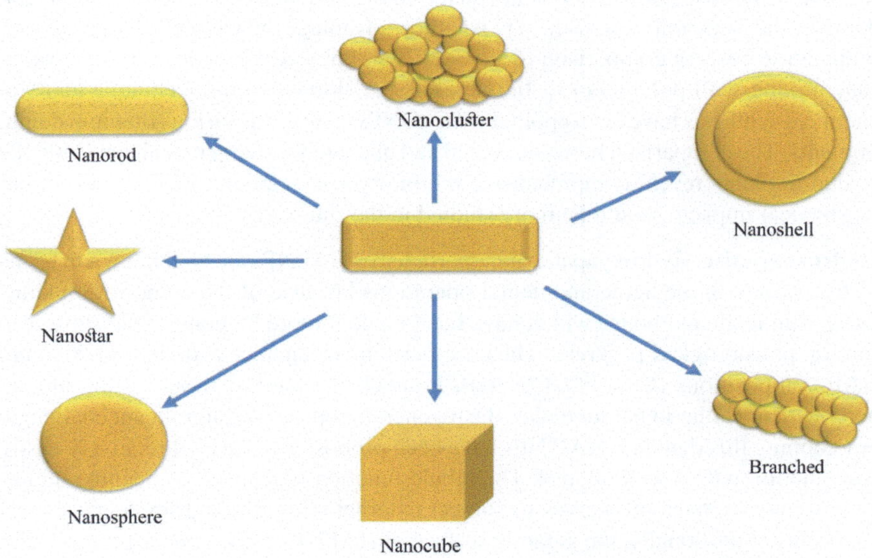

Fig. 5.2 Different shapes of gold nanoparticles (Reprinted from J. Environ. Anal. Toxicol, Alaqad et al., Gold and silver nanoparticles: synthesis methods, characterization routes and applications towards drugs, no. 2, Copyright (2016), with permission from OMICS international, CC BY 4.0) [12]

due to the exopolymeric matrix of microorganisms in the biofilm, which is impenetrable by most antibiotics and host immune cells. But SPIONs have trodden a path to overcome the problem because of relatively easy penetration of biofilms. Various portions of iron oxide NPs on different biomaterials surfaces have fluctuating effects on the biofilm formation; it was reported that polymer brush-coated surfaces decrease bacterial adhesion by comparison to the bare surfaces. The noteworthy reduction was reported for biofilm growth of S. aureus, Escherichia coli, and P. aeruginosa in the presence of iron oxide NPs. Additionally, iron oxide NPs have magnetic properties which are relatively stable and safe in comparison to the other magnetic NPs at which the addition of silica enriches the stability and functionality. Cytotoxicity test by an MTT assay (3-(4,5-dimethylthiazol-2-yl)-2,5-diphenyltetrazolium boride) for diverse sizes range with varied concentrations of nanocomposite has been done. The outcomes reveal that iron oxide NPs size had minor impact on fibroblast viability, although an NP concentration pf 1000 μg/ml regardless of nanocomposite size range reduced macrophage viability. So, in spite of the size range, the concentration has an effect on cellular toxicity [1].

Copper Oxide. Copper oxide (CuO) nanoparticles or CuONPs are common due to the physicochemical stabilities and antibacterial actions similar to the silver operations. In spite of that, CuONPs exhibit effective antibacterial activities in high concentration which could lead to be toxic to mammalian cells [3]. Generation of reactive oxygen species (ROS) is attributed to the antibacterial activities which can damage the bacterial cell wall [1]. Another advantage of CuONPs is usage and preparation costs in comparison to silver NPs and physical properties that allow an easy mixing with polymers [3]. In one case, the slurry dipping technique and wet chemical synthesis have been applied for CuONPs coating on various titanium dental implants. The comparison between coated and uncoated surfaces reveal that CuONPs coated implants reveal comprehensive inhibitory zones around 31.6–38.6 mm but mechanical impacts should be more studied in the future [1].

Hydroxyapatite. Hydroxyapatite nanoparticles or HAPNPs are well known in terms of their usage in medicine and dental operations because of the composition similarity with teeth and bones which have been made it more biocompatible substance for any physiological progress. The core portions of enamel with around 97% are HAP compositions ($Ca_{10}(PO_4).2(OH)$). Enamel, the hardest tissue of the human body with acellular in nature makes the regeneration process a significant challenge. For copping this demand, HAP NPs have been proposed because of relatively easily combination with dental tubules. The tubule function is sealing the opening and as a consequence, bans the nerves to subject external stimuli and drop dental hypersensitivity. Additionally, the gigantic surface of HAPNPs causes powerful bind with proteins as same as with bacterial and plaque fragments, and also a high biological activity which provided them for strong bonds to the dentin apatite and tooth enamel. Nanoparticle scale of HAP is attributed to sturdy absorption to the enamel and thus slowing down auxiliary erosive demineralization. In terms of biocompatibility, HAP is capable of attaching to the bone and does not show any locally, systemic, or toxic phenomena inflammatory-causing [6]. It was reported that HAPNPs can be prepared by different synthesis manners which are divided into solid state or wet state. The modifications could be done by ion substitution using cations such as Mg^{2+}, $Sr^{2+,}$ Zn^{2+}, Fe^{3+} and Ag^{+}, and/or anions such as $CO_3{}^{2-}$, $SiO_4{}^{3-}$ and F^- to revise their biocompatibility, sintering ability, and mechanical properties. Likewise, modification via absorption of organic molecules on HAP surfaces and additives can be applied [1]. Table 5.6 presents different preparation methods for HAPNPs introduced which are dry and wet methods, high-temperature processes, synthesis from biogenic sources, and combination procedures.

Titanium Dioxide. Titanium dioxide nanoparticles or TiO_2 NPs are widely used in medical and dentistry [6]. The antibacterial action mechanism of TiO_2 NPs is attributed to the hydroxyl radicals formation by photocatalytic reactions in water environment which destroys polyunsaturated phospholipids in the peptidoglycan cell membrane and, consequently, attacking the cell and spoiling cell DNA [1]. The application of implants leads to the allergic-induced activities by antigen/antibody type I and IV complex reaction. It was highlighted that roughness and chemical decomposition of implant surfaces could form plaque inside the oral cavity and also adhesion of microbes on implants has a major impact on teeth curing process. As a consequence, modification via stable titanium dioxide (titanium/zirconia NPs) reduces the adhesion of bacterial compared to free polished titanium [6]. Addition of TiO_2 NPs to conventional glass ionomer (GI) enhanced properties without affecting GI roles (i.e. bond strength and flowing out fluoride) which could be used for the restoration of class I and II caries with higher durability of stress-bearing regions. Likewise, TiO_2 NPs could be used in dental resins for agglomeration controlling and composites for improvement in mechanical properties [1].

Calcium Fluoride. The optimized concentrations of fluoride assist to control the formation of caries, at which in one approach, the modification of restorative materials with fluoride compounds like CaF_2 occurred which could flow out fluoride ions and provide remineralization of dental tissue and nearby the restoration. This combination of fluoride is in the aim of enhancement of ions activities. In a different research, it was underlined that CaF_2 nanoparticles or CaF_2 NPs prove strong antibacterial activity against S. mutans that leads to a reduction around 90% of biofilm formation, reduced bacterial acid, and exopolysaccharides productions. The bacterial cell death happens when pH is low and fluoride and hydrogen ions bind create hydrofluoric acid (HF) which penetrates into bacterial membrane, leads acidification of cytoplasm and enzymes inhibition (enolase and ATPase) and finally, in optimum concentration (3040–5700 ppm), causes cell death. More than optimum concentration, bacterial cells adhesion to tooth surface decrease. Because of great fluoride ion release and its effect on bacteria, CaF_2 NPs stop biofilm formation and drop caries lesion development. Besides, calcium plays a role as a remineralization agent which is the enamel structure element [14]. So, the combination of calcium and fluoride in nanoscale is more productive which can act as a fluoride reservoir.

Calcium Phosphate. Calcium phosphate ($CaPO_4$) nanoparticles or CPNPs have been proposed for fulfilling the limitation of Ca and PO_4 penetration into the inner

Table 5.6 Preparation methods for HAP NPs by definition, advantages, and disadvantages [3]

Preparation Method	Definition
Dry methods	Using solvent in solid-state synthesis and mechanochemical process **Advantages**: Producing high crystalline HAP via inexpensive raw materials, **Disadvantages**: Low phase purity and large size of particles in the mechanochemical process and solid-state synthesis, respectively
Wet methods	Precipitation of HA by using an aqueous solution of phosphate and calcium ions sources which divided into conventional chemical precipitation, hydrolysis, sol–gel, hydrothermal, emulsion, and sonochemical methods **Advantages**: A promising method for nanosized synthesis, controlling over the morphology **Disadvantages**: Difficulty in phase purity and crystallinity controlling, time-consuming, not completely suitable for scaling up
High-temperature processes	Two main processes for HAP are Combustion and pyrolysis method **Advantages**: Avoiding undesirable calcium phosphate phases, homogenous, and high crystallinity HAP **Disadvantages**: Poor control over the processing factors, secondary aggregates creation
Synthesis from biogenic sources	Raw materials are from nature like a bone waste, eggshells, the exoskeleton of marine organisms, bio-membranes, etc. **Advantages**: Better physicochemical properties of HAP **Disadvantages**: May produce large size particles (need more attention to study)
Combination procedures	Employment of two or more distinct procedures to synthesis HAPNPs such as hydrothermal/microemulsion and microwave/hybrids **Advantages**: Improve the characteristics of HAPNPs (need more attention to study)

layer of the affected enamel and could treat affected dentin tissues. The nanosize of particles provided most release of Ca and PO_4 ions which not only do enhances the treatment procedure but also improves mechanical properties [1].

Silica. Silica nanoparticles or SiNPs play an important role in nanoscience due to the special properties like size, surface area portion, biocompatibility, little toxic potential, and adsorption capability which have been used in drug delivery systems, enzyme supporters and dentistry as a filler for mechanical properties enhancement [6]. Applied dental composites are usually modified by inorganic fillers and organic matrix which could be divided into three main types based on silica usage: filler-free (direct filling), macrofiller via large particles of glass and micro filler with the small size of silica particles systems [15]. Additionally, SiNPs are consumed as a dental polishing agent for biocompatibilities and reasonable cost. This protecting process is done due to acid-causing agents like cariogenic bacterial which could increase the demineralization of dental surfaces and producing caries and known as an initial protection stage against cariogenic bacteria [6].

Chitosan. Chitosan nanoparticles CSNPs are organic materials known as antibacterial components for dental application. The antibacterial activity accounts for the chemical structure which deacetylated C2 amino groups change into positive charge by protonation at pH <6.5 and consequently, chitosan binds to the bacterial cells membrane which causes growth in permeability of membrane with attendant increase in outward flow of ions and proteins from cell and inhibition of mRNA transcription and changing of proteins translation which is binding chitosan to the DNA of cells. Additionally, the molecular weight of CSNPs is directly related to the antibacterial actions which account for the destabilization of the bacterial membrane. The stabilization of the outer membrane of bacterial cells happens by Mg^{2+} and Ca^{2+} ions that if replaced by CSNPs, modification of cell wall progressed in parallel via oscillation of degradative enzymes activities. Primary enamel carious lesions would be demineralized via CSNPs [16].

Zirconia. Zirconia nanoparticles or ZrNPs have a considerable application in medical science especially in dentistry because of impressive property causing for host materials. For instance, in denture base resins, it can enhance mechanical properties which result from an increase in distribution and silanization of ZrNPs in applied restorative material. These NPs provide compressive stress due to the expansion of ZrO_2 crystals and consequently, arrest crack propagation in the dentine. Likewise, ZrNPs in implants could shield against corrosion and carry. The similarity in color with the tooth is another important property of ZrNPs that affects the application, positively. Additionally, it is not a water-soluble substance that drops bacterial adhesion and provides low cytotoxicity. Osseo conductivity of the ZrNPs is productive for the bone formation process. The ZrNPs is incorporated by different materials like alumina and titania. The former combinations lead to toughness and longevity enhancement and the latter formation increase in compressive, tensile, and shear bond strength [1, 6].

Zinc Oxide. Zinc oxide nanoparticles or ZnONPs are other functional, and versatile inorganic materials. Different researchers have been examined the antibacterial action of ZnONPs at which the proposed mechanisms of action are listed as the direct contact of ZnONPs with bacterial cell walls, the liberation of antimicrobial ions mainly Zn^{2+} ions and as a result ROS formation [17]. Application of ZnONPs in composite resins could affect antibacterial activities like inhibition of biofilm formation, in a positive way with mentioned mechanisms of action [1].

PLGA. Poly (lactic-co-glycolic acid) or PLGA is a kind of biodegradable polymer which was synthesized and used successfully, with the aim of medical application and has been accepted by the US Food and Drugs Administration and European Medicine Agency. The reasons for PLGA popularity in medicine are biocompatibility, biodegradability, flexibility, and minimal side effects. Additionally, different degradation actions of PLGA are described as hydrolysis, oxidation, and enzymatic reactions which are discussed more in Fig. 5.3.

PLGA nanoparticles or PLGA NPs are well developed for medical applications like gene carriers, vaccine delivery systems, and in the anticancer treatment. Likewise, it was developed for dental applications that are presented in Table 5.7. For instance, the minocycline-loaded PLGA NPs did not only show productive antibacterial activities but also suggested a potential carrier system for antibiotics delivery to periodontal tissues [18].

Carbon Nanotube (CNT). As can be seen in Fig. 5.4, carbon nanotubes or CNTs are made by a rolled-up graphene sheet via hemispherical, half-fullerene endcaps which can be divided into single- or multi-walled carbon nanotubes. The former is also called SWCNTs and has one graphene sheet but the latter involves many graphene sheets rolled-up in one another and is called MWCNTs. These materials show impressive mechanical properties because of the strength of their carbon sp^2 bonds and are directly related to the preparation methods. Plus, CNTs have thermal and electrical conductivity, which alongside the excellent mechanical properties,

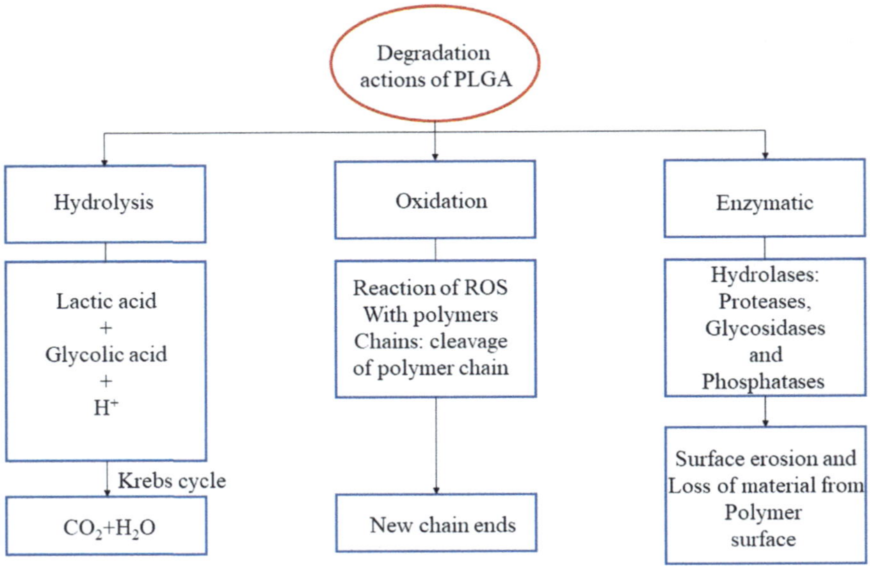

Fig. 5.3 Different degradation actions of PLGA that are hydrolysis, oxidation, and enzymatic processes. Redrawn from Virlan et al. [18]

Table 5.7 PLGA NPs application in dentistry (Reprinted from Journal of Chemistry, Virlan et al., Current uses of poly (lactic-co-glycolic acid) in the dental field: a comprehensive review, no.7, Copyright (2015), with permission from Hindawi, CC BY 3.0) [18]

Type of PLGA nanoparticles	Incorporation in	Dental applications
PLGA NPs	Minocycline	Periodontal infections
Dental implants coated with PLGA NPs	Basic fibroblast growth factor	Implantology
PLGA NPs	Methylene-blue photosensitizer	Endodontic infections
PLGA NPs	Methylene-blue photosensitizer	Periodontology (reduction of dental plaque biofilms)
PLGA NPs heparin-conjugated	BMP-2 (bone morphogenetic protein-2)	Bone regeneration (osteogenic differentiation of bone marrow stem cells)
PLGA NPs	Simvastatin	Bone regeneration (enhanced osteogenesis of bone marrow mesenchymal stem cells)

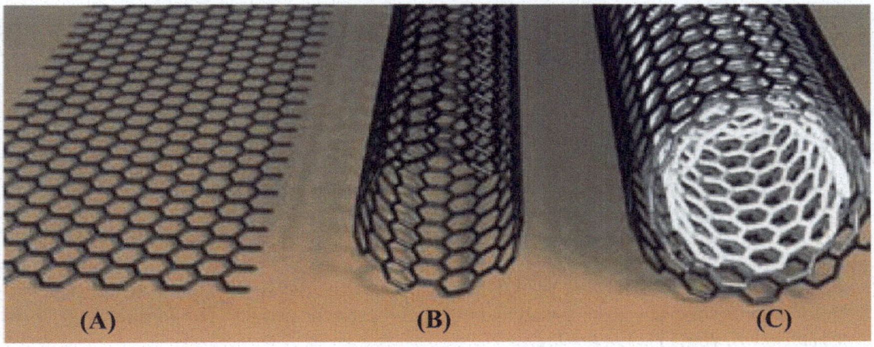

Fig. 5.4 Schematic of **A** graphene sheet, **B** single-walled carbon nanotube (SWCNT), and **C** multi-walled carbon nanotube (MWCNT), (Reprinted from Applied Physics A, Graham et al., How do carbon nanotubes fit into the semiconductor roadmap? no. 1141, Copyright (2005), with permission from Springer) [20]

provide a good choice for dental filling and wide applications. About preparation of CNTs, three methods have been used which are arc discharge, laser ablation, and chemical vapor deposition (CVD). In the arc discharge methods, the CNTs are sterile and pristine with many nanoparticles and have a yield of around 2% after purification. In the second method, laser ablation could prepare top-quality CNTs and the yield is as well as the first method. In the CVD method, the most popular and common one, the CNTs are provided inexpensively in large production. By increasing the reaction temperature, the products would have a higher degree of graphitization and the fewer defects in their structure. However, many defects like topological defects, rehybridization, and incomplete bonding are inevitable which impact the mechanical integrity of the CNTs [1, 19].

Graphene. Graphene, a monolayer of carbon atoms, is tightly, closely packed into 2D crystals which contain sp^2 hybridized carbon atoms and is also known as the thinnest material with the uniform crystal lattice, vacancy-free by without any liner defects like dislocations. For these, graphene got conducting electrons which are attributed for usage in photovoltaics, biodevices, and producing antibacterial surfaces. Graphene could be applied with different materials such as zinc oxide for preparation of a nanocomposite which is capable of inhibiting biofilm formation via streptococcus mutans. Additionally, it is used as a coating for acrylic teeth with low-cost, fracture resistance, and suitable density advantageous [6, 21].

References

1. Bapat, R. A., Joshi, C. P., Bapat, P., Chaubal, T. V., Pandurangappa, R., Jnanendrappa, N., … & Kesharwani, P. (2018). The use of nanoparticles as biomaterials in dentistry. *Drug discovery today*.
2. Mantri, S. S., & Mantri, S. P. (2013). The nano era in dentistry. *Journal of Natural Science, Biology, and Medicine, 4*(1), 39.
3. Chaughule, R. S. (Ed.). (2018). *Dental Applications of Nanotechnology*. Springer.
4. Freitas, R. A., Jr. (2000). Nanodentistry. *The Journal of the American Dental Association, 131*(11), 1559–1565.
5. Akbarianrad, N., Mohammadian, F., Alhuyi Nazari, M., & Rahbani Nobar, B. (2018). Applications of nanotechnology in endodontic: A review. *Nanomedicine Journal, 5*(3), 121–126.
6. Priyadarsini, S., Mukherjee, S., & Mishra, M. (2018). Nanoparticles used in dentistry: A review. *Journal of oral biology and craniofacial research, 8*(1), 58–67.
7. Peng, T., Zhao, D., Dai, K., Shi, W., & Hirao, K. (2005). Synthesis of titanium dioxide nanoparticles with mesoporous anatase wall and high photocatalytic activity. *The Journal of Physical Chemistry B, 109*(11), 4947–4952.
8. Hu, Y., Jiang, X., Ding, Y., Ge, H., Yuan, Y., & Yang, C. (2002). Synthesis and characterization of chitosan–poly (acrylic acid) nanoparticles. *Biomaterials, 23*(15), 3193–3201.
9. Kartick, B., & Srivastava, S. K. (2013). Green synthesis of graphene. *Journal of Nanoscience and Nanotechnology, 13*(6), 4320–4324.
10. Bapat, R. A., Chaubal, T. V., Joshi, C. P., Bapat, P. R., Choudhury, H., Pandey, M., … & Kesharwani, P. (2018). An overview of application of silver nanoparticles for biomaterials in dentistry. *Materials Science and Engineering: C, 91,* 881–898.
11. Chen, P. C., Mwakwari, S. C., & Oyelere, A. K. (2008). Gold nanoparticles: from nanomedicine to nanosensing. *Nanotechnology, Science and Applications, 1,* 45.
12. Alaqad, K., & Saleh, T. A. (2016). Gold and silver nanoparticles: Synthesis methods, characterization routes and applications towards drugs. *Journal of Environmental & Analytical Toxicology, 6*(4), 525–2161.
13. Jadhav, K., Rajeshwari, H. R., Deshpande, S., Jagwani, S., Dhamecha, D., Jalalpure, S., … & Baheti, D. (2018). Phytosynthesis of gold nanoparticles: Characterization, biocompatibility, and evaluation of its osteoinductive potential for application in implant dentistry. *Materials Science and Engineering: C, 93,* 664–670.
14. Łukomska-Szymańska, M., Kleczewska, J., Nowak, J., Pryliński, M., Szczesio, A., Podlewska, M., … & Łapińska, B. (2016). Mechanical properties of calcium fluoride-based composite materials. *BioMed research international*.

15. Kim, J. W., Kim, L. U., & Kim, C. K. (2007). Size control of silica nanoparticles and their surface treatment for fabrication of dental nanocomposites. *Biomacromolecules, 8*(1), 215–222.
16. Padovani, G. C., Feitosa, V. P., Sauro, S., Tay, F. R., Durán, G., Paula, A. J., et al. (2015). Advances in dental materials through nanotechnology: Facts, perspectives and toxicological aspects. *Trends in Biotechnology, 33*(11), 621–636.
17. Sirelkhatim, A., Mahmud, S., Seeni, A., Kaus, N. H. M., Ann, L. C., Bakhori, S. K. M., ... & Mohamad, D. (2015). Review on zinc oxide nanoparticles: Antibacterial activity and toxicity mechanism. *Nano-Micro Letters, 7*(3), 219–242.
18. Virlan, M. J. R., Miricescu, D., Totan, A., Greabu, M., Tanase, C., Sabliov, C. M., ... & Calenic, B. (2015). Current uses of poly (lactic-co-glycolic acid) in the dental field: A comprehensive review. *Journal of Chemistry.*
19. White, A. A., Best, S. M., & Kinloch, I. A. (2007). Hydroxyapatite–carbon nanotube composites for biomedical applications: A review. *International Journal of Applied Ceramic Technology, 4*(1), 1–13.
20. Graham, A. P., Duesberg, G. S., Hoenlein, W., Kreupl, F., Liebau, M., Martin, R., ... & Unger, E. (2005). How do carbon nanotubes fit into the semiconductor roadmap?. *Applied Physics A, 80*(6), 1141–1151.
21. Hu, W., Peng, C., Luo, W., Lv, M., Li, X., Li, D., ... & Fan, C. (2010). Graphene-based antibacterial paper. *ACS nano, 4*(7), 4317–4323.

Chapter 6
3D-Printing Technologies for Dental Material Processing

6.1 Introduction

Additive manufacturing (AM) especially the three-dimensional (3D) printing technology has been applied for creating different complex geometry constructions which consists of printing successive layers of materials on top of other layers. Charles Hull in 1986 firstly introduced the 3D-printing method in a process that he so-called stereolithography (SLA). 3D printing offers material bodies with a minimum requirement for post-processing, around zero material waste and the capability of wide application of starting materials such as ceramics, metals, and polymers. Also this technology provides a free-design for manufacturing complex bodies. Other advantages like high efficiency in materials saving, personal customization, minimum additional expenses for developing a product, and rapid production cause this technology to be more attractive. Therefore, like any processing methods, 3D-printing technology has different limitations. For instance, product-customization is still a problem for manufacturer due to the expensive production for custom-offered productions. But it was predicted that in the future, more than 50% of customized functional products become fashion and trend. In addition, one of the active fields in 3D-printing products is medical application such as implants and dental materials which are presented in the following subsections [1, 2].

3D-printing manner has different techniques (Fig. 6.1), which are selective laser sintering (SLS), selective laser melting (SLM), direct metal laser sintering (DMLS), and so forth. In this chapter, the most important ones are reviewed, see Tables 6.1 and 6.2 [3, 4].

H. Reza Rezaie et al., *A Review on Dental Materials*, Advanced Structured Materials 123, https://doi.org/10.1007/978-3-030-48931-1_6

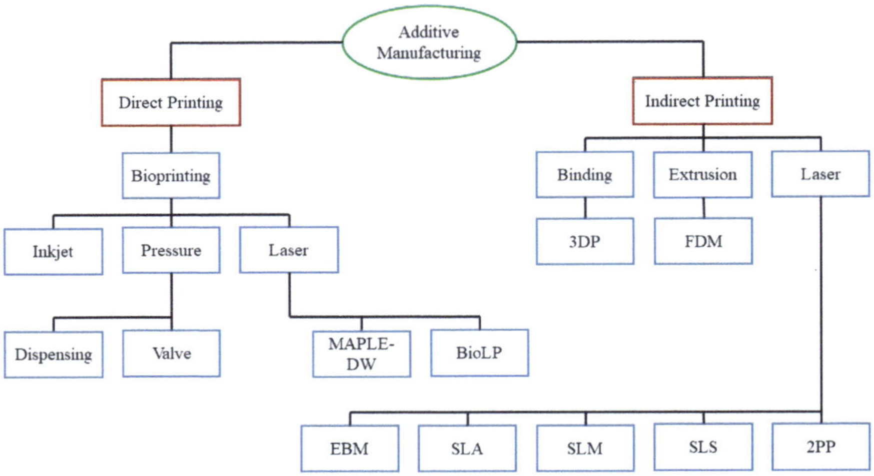

Fig. 6.1 An overview of additive manufacturing (AM) techniques (Modified and Reprinted from International Journal of Precision Engineering and Manufacturing, Ho et al., A review on 3D printed bioimplants, no. 1036, Copyright (2015), with permission from Springer) [10]

6.2 Different 3D-Printing Techniques

Selective Laser Sintering (SLS). One of the biggest applied AM processes is SLS due to the high degree of accuracy and surface quality. The starting powders of SLS are fine powders at which a laser beam plays the heating role for fusing particles together by changing the power between 7 and 200 W that the former is for plastics. Before applying the laser beam, the entire bed is heated up to the melting point of the powder for minimizing thermal distortion and facilitate fusion in the previous layers. Then, new powder layers are used when each layer was made and the bed is lowered. Subsequently, a rotating wheel is employed with the aim of making a uniform powder spread. In parallel, the sintered materials are forming the body, unsintered materials are existing to hold the structures which are recyclable after processing [1]. The laser source could be CO_2 or Nd:YAG radiation. The starting powders involve four types, i.e. polymers (e.g. polyamides, polystyrenes, polycarbonates and nylon), ceramics (e.g. HA, glasses, and glass-ceramics), metals (e.g. steel, titanium, aluminum, and silver) and composites (e.g. HA/PEEK) [4].

Selective Laser Melting (SLM). Most portion of the SLM process is similar to the SLS process, but instead of sintering, powder melting occurs in order to form the parts and consequently, the laser power is higher, i.e. around 400 W [1].

Direct Metal Laser Sintering (DMLS). Similar to the SLS/SLM, this process is well working. However, DMLS has a limitation in regards to the starting material, which is just metallic powder. Additionally, the laser powder is altering from 200 W to 400 W. This technology is developed by EOS GmbH of Munich, Germany and is

Table 6.1 Some commercial AM techniques in terms of manufacturers and material providers (Reprinted from Journal of Manufacturing Processes, Singh et al., Material issues in additive manufacturing: A review, no. 186, Copyright (2017), with permission from Elsevier) [3]

AM technique	Manufacturers	Material providers
SLS, selective heat sintering	3DSYSTEMS, Farsoon Technologies, Blueprinter (Denmark), eos, MC Machinary Systems, Wuhan Binhu Mechanical & Electrical Co., Ltd	3DSYSTEMS, Paramount Product Development Specialists, Farsoon Technologies, Concept Laser, eos, LuxExcel, Wuhan Binhu Mechanical & Electrical Co., Ltd
Selective laser melting (SLM)	MICROFABRICA, REALIZER, SLM Solutions, Wuhan Binhu Mechanical & Electrical Co., Ltd	MICROFABRICA, REALIZER, SLM Solutions, Wuhan Binhu Mechanical & Electrical Co., Ltd
Metal laser melting (MLM)	Concept Laser	Concept Laser
EBM	Arcam (Sweedon), Sciaky	Arcam (Sweden), Sciaky
Laser metal deposition (LMD)	BEAM (France)	BEAM (France)
Inkjet, poly jet, binder jet	Envisiontec, ExOne, MicroFab, Stratasys, Solidscape	Envisiontec, ExOne, MicroFab, Prodways, Stratasys, Solidscape
SLA, lithography-based ceramic manufacturing	3DSYSTEMS, CMET Inc., Somos, Lithoz, Prodways, Wuhan Binhu Mechanical & Electrical Co., Ltd	3DSYSTEMS, CMET Inc., Somos, Lithoz, Prodways, Wuhan Binhu Mechanical & Electrical Co., Ltd
Digital wax machine	DWS	DWS
FDM	UP-3D, Stratasys, CandyFab, Fab@Home, MAKERBOT, Solidscape, Polyflex	UP-3D, Stratasys, CandyFab, Fab@Home, MAKERBOT, LuxExcel, RepRap, Solidscape, Argyle Materials Inc., Polymakers®
DMLS	Insstek	Concept Laser, Insstek
Metal powder bed fusion (MPBF)	Renishaw	Renishaw
Ultrasonic additive manufacturing (UAM)	Fabrisonic	Fabrisonic
LENS	Optomec	Optomec, Concept Laser
LOM	Cubic Technologies	Cubic Technologies
3DP	Asiga, Keyence Corporation	Asiga, Keyence Corporation
ZCast	ZCast	ZCast

commercially well known as EOSINT M 250 laser sintering machine. The EOS is two powder systems of bronze based and steel based [5].

Selective Laser Cladding (SLC). The laser is used to melt a metal powder for deposition on a substrate. SLC is a layer-by-layer and point-by-point technique to form functional metal parts with near-net shape geometries. SLC processing has

Table 6.2 Family of additive manufacturing processes (Modified and reprinted from The International Journal of Advanced Manufacturing Technology, Bikas et al., Additive manufacturing methods and modelling approaches: a critical review, no.395, Copyright (2016) with permission from Springer) [1]

Additive manufacturing (AM) processing								
Process		Laser-based AM process			Extrusion thermal	Material jetting	Material adhesion	Electron beam
		Laser melting		Laser polymerization				
Process Schematic		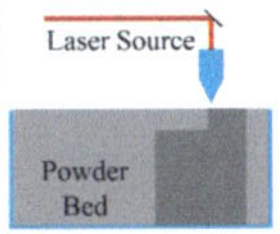	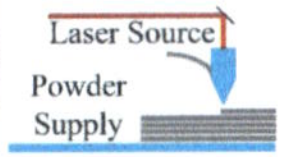	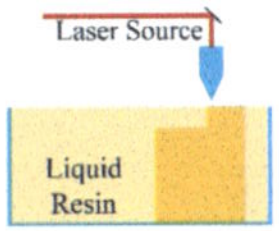		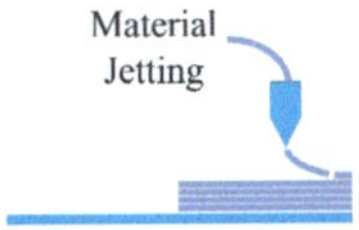	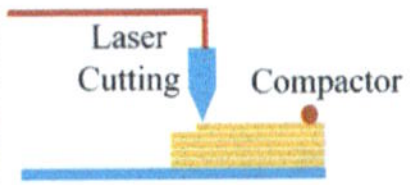	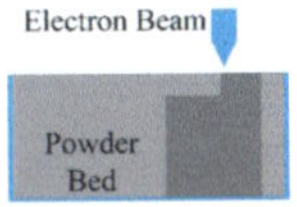
Name	Material	SLS	DMD	SLA	FDM	3DP	LOM	EBM
		SLM	LENS	SGC	Robocasting	IJP	SFP	
		DMLS	SLC	LTP		MJM		
			LPD	BIS		BPM		
				HIS		Thermojet		
Starting material type			Powder	Liquid	Solid			

Remarks:
SLS (selective laser sintering), SLM (selective laser melting), DMLS (direct metal laser sintering), DMD (direct metal deposition), LENS (laser engineered net shaping), SLC (selective laser cladding), LPD (laser powder deposition), SLA (stereolithography), SGC (solid ground curing), LTP (liquid thermal polymerization), BIS (beam interference solidification), HIS (holographic interference solidification), FDM (fused deposition modeling), 3DP (three-dimensional printing), IJP (inkjet printing), MJM (multijet modeling), BPM (ballistic particle manufacturing), LOM (laminated object manufacturing), SFP (solid foil polymerization), EBM (electron beam melting)

been overweight of conventional rapid manufacturing (RM) methods because of direct fabrication of a near-net shape part [1].

Laser Engineered Net Shaping (LENS). Metal powders are melted by a high-power laser source. A nozzle injects the powder stream directly into the focused laser beam at which the laser head and powder nozzle are integral parts in movement. Plus, metal powders are spread around the circumference of the head by its weight and/or pressurized carrier gas. Then, a small molten batch of metal powders is formed on the previously deposited layer or on the substrate. Finally, the powder is consumed in this paddle which is attributed to its height to grow away from the substrate surface, then an x-y table is moved to fabricate the layers of the object, afterward, the head is moved up vertically after each layer is finished. This technique is overlapped with DMD, LPD, and SLC methods since it creates larger bodies in the order of several meters, but the accuracy and quality are weaker than that of the mentioned techniques [1].

Stereolithography (SLA). This technique uses a photopolymerization curing agent (e.g. ultraviolet (UV) light or digital light projection (DLP)) for photosensitive materials like low molecular weight polyacrylates, epoxy macromers, or monomers. In this technique, after specifying the interested locations for the aim of subjection to the light by a digital micro-mirror device, the light radiation starts the chemical reaction of gelation. This process is continued until all the layers are completed and the final expected object is prepared. However, due to the photosensitivity of the resins, the long-service and stability of the products are not satisfying [6].

Electron Beam Melting (EBM). This process is a relatively state-of-the-art AM technique which is a rapidly growing process and suitable for production of metallic bodies. In EBM, powder is melted via an electron beam which is powered by voltage in order of 30–60 kV. The progress is under a vacuum condition in order to keep away the products from any oxidation. In comparison with the SLS technique, EBM could provide higher number of products with uniform heat affected fields. By the way, it has a poor accuracy and surface quality [1].

Semisolid Extrusion (EXT). This technique is well known as the pressure-assisted microsyringe (PAM) additive manufacturing technique which is based on extrusion of semisolid materials such as gels or pastes through the nozzle on a prepared plate. By starting the first layer deposition on the plate, the plate is lowered and this repeated cycle is taken place till the final body is completed. EXT technique was named extrusion-based bioprinting technique after the first usage for biomedical application (e.g. printing living cell and different biologics for printing living organs). However, the shrinkage and structure deformation issues happen in final products due to the semisolid form of starting materials (i.e. pastes and gels) [4].

6.2.1 3D-Printing Application in Oral Cavity Restorative Material

As it is mentioned before, 3D-printing technologies have a critical role in different areas especially in biomedical applications. For instance, they have a wide application in the production of complex organs, ophthalmology, drug delivery, etc. In recent years, 3D-printing techniques have various applications in dentistry in terms of making different dental materials, total or partial replacement of alveolar bone, replacement of damaged blood vessels, nerves, muscles, and so on [7, 8]. Different range of dental materials (i.e. metals, ceramics, and polymers) that could be produced via 3D-printing process are indicated in Table 6.3. These materials are bioinert, corrosion resistance, high volume to mass ratio, bioresorbable, and so forth. Moreover, in Table 6.4, different 3D-priniting technologies for dentistry based on localized oral cavity applications are mentioned.

As an example, in some dental treatment, partial or whole portion of the jaw should be replaced due to cancer spread. In this case, 3D-printing techniques (e.g. SLM) are applicable for producing the infected region with titanium powders (Fig. 6.2). The bioinert part will not be affected by body fluids. In other cases, bioresorbable maxillary implants with calcium phosphate component are made via dissolving the material

Table 6.3 3D-printing technologies based on the type of materials (Reprinted from Journal of Biomedical Materials Research Part B: Applied Biomaterials, Bhargav et al., Applications of additive manufacturing in dentistry: A review, no. 4, Copyright (2018), with permission from John Wiley and Sons) [8]

Family of materials	Materials	Advantages	3D-Printing techniques
Metals	Titanium	Bioinert, corrosion resistance	Selective laser sintering, direct metal
	Stainless steel	Bioinert, corrosion resistance	Laser sintering, selective laser melting
	Aluminum	High volume to mass ratio	Laser sintering, selective laser melting
	Cobalt chrome	Corrosion resistance	Direct metal deposition
Ceramics	Alumina	Bioinert, same color as tooth, corrosion resistance	Selective laser sintering, stereolithography, inkjet 3D
	Zirconia	Bioinert, same color as tooth, corrosion resistance	Printing, thermoplastic printing
Polymers	Poly lactide	Bioresorbable, odontogenic	Stereolithography, fuse deposition
	Poly caprolactone	Bioresorbable	Modeling, electro hydrodynamic
	Poly glycolide	Bioresorbable	Jetting
	Acrylic	Bioresorbable	Extrusion based

Table 6.4 Oral cavity materials made by 3D-printing techniques (Reprinted from Journal of Biomedical Materials Research Part B: Applied Biomaterials, Bhargav et al., Applications of additive manufacturing in dentistry: A review, no. 4, Copyright (2018), with permission from John Wiley and Sons) [8]

Oral cavity application	Materials	3D-Printing techniques	Positive outcomes
Maxillo-facial implants	Titanium	Selective laser sintering, direct metal deposition	Bioinert
	Calcium phosphate	Photopolymerization, inkjet 3D printing	Bioactive or bioresorbable, mechanical compliance similar to tooth
Dentures and crowns	Cobalt chrome	Selective laser sintering	Corrosion resistant
	Plastic	Fuse deposition modeling, inkjet 3D printing	Ease of fabrication
	Alumina	Thermoplastic printing	Bioinert
	Zirconia	Direct inkjet printing, thermoplastic printing	Bioinert
Anatomical and training models	Metals	Selective laser sintering, direct metal laser sintering, selective laser melting, direct metal deposition	Sturdy models
	Veroglaze	Polyjet 3D printing	Same color as tooth
	Nylon	Stereolithography, selective laser Sintering	Same color as tooth
	Plaster	Indirect 3D Printing, Inkjet Printing	Easy of fabrication, inexpensive
Scaffolds for tissue engineering	Poly lactide	Fused deposition modeling, electrospinning, stereolithography	Ease of fabrication, bioresorbable, stereolithography
	Poly Caprolactone	E-jetting, electrospinning	Ease of fabrication
	Ceramics	Selective laser sintering, polyjet printing	Same color as teeth, good mechanical properties
Scaffold-less tissue engineering	Cells	Microextrusion	Eliminates need for Immunosuppressant

into a photocurable mix and subsequently subjected to a suitable light for bounding of layer to each other. According to the previous chapters, calcium phosphates are present in the body and will increase the biocompatibility of the implant [8].

3D printing is also used for denture replacement. Different factors are involved in damaging the denture which cause removing all teeth or the complete denture.

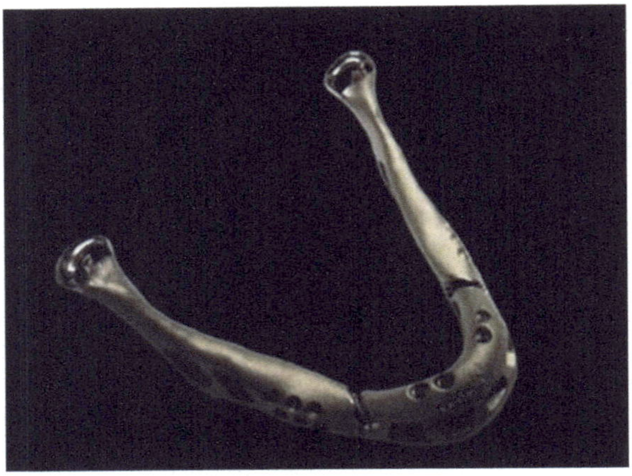

Fig. 6.2 Man-made jaw via 3D-printing technology (Reprinted from Journal of Biomedical Materials Research Part B: Applied Biomaterials, Bhargav et al., Applications of additive manufacturing in dentistry: A review, no. 5, Copyright (2018), with permission from John Wiley and Sons) [8]

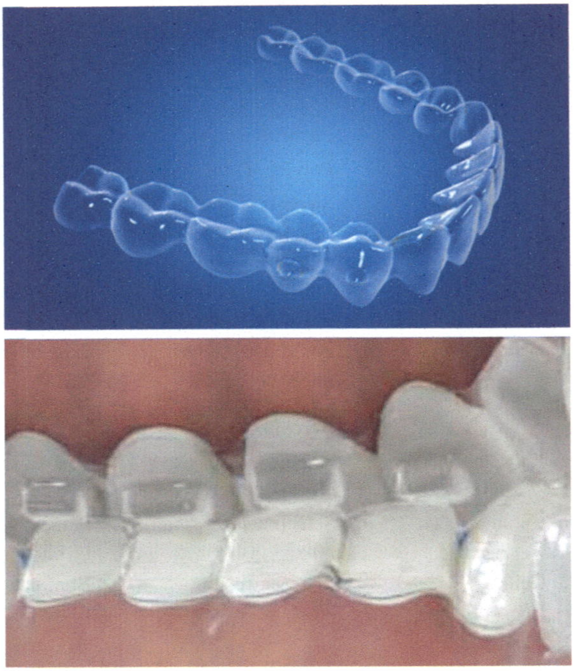

Fig. 6.3 Transparent aligners which were produced via SLA techniques (Reprinted from Chemical Reviews, Ligon et al., polymers for 3D printing and customized additive manufacturing, no. 10252, Copyright (2017), with permission from American Chemical Society) [9]

Metallic denture could be manufactured via DMLS technique with composition of cobalt, chromium, and molybdenum elements. Likewise, polymers and plastics are used for temporary implant parts, which are produced by 3D-printing techniques (e.g. SLA), too [8]. For example, dental brackets are possible to be made by SLA and inkjet printing for an alternative to the metal brackets (Fig. 6.3). These thin transparent plastic inserts placed incrementally every two weeks on top of the user's teeth. These aligners could be manufactured by SLA technique through a polymeric mold with a thickness of 1 mm [9].

Dental crowns are another part that can be prepared via SLS or SLM technique based on the crown type (Fig. 6.4) [8].

Moreover, 3D-printing techniques could be employed in dental tissue engineering (i.e. the combination of cells and substances for fabricating native tissues, and the cells which are creating a formation of 3D-structure are called scaffold) fields and this research area is still state of the art. In pulp cavity filling materials, the structure can be prepared via 3D printing that cells are formed a preprogrammed structure. As a consequence, there is no need for any scaffold tissue. For instance, hydrogels are delivering cell deposition which bioinks comprising cells, media, and different 3D-printing materials [8].

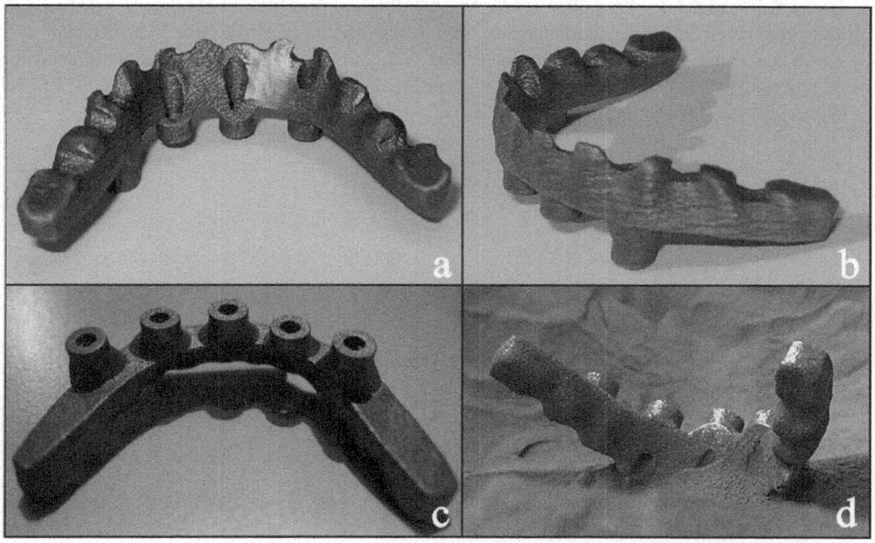

Fig. 6.4 Prepared frameworks via 3D-printing techniques of SLS/SLM. **a**, **b** made from stainless steels, **c**, **d** made from Ti6Al4V alloys. The last part (**c**) illustrates an early state of framework in powder bed (Reprinted from International Journal of Precision Engineering and Manufacturing, Ho et al., A review on 3D printed bioimplants, no.1040, Copyright (2015), with permission from Springer) [10]

References

1. Ho, C. M. B., Ng, S. H., & Yoon, Y. J. (2015). A review on 3D printed bioimplants. *International Journal of Precision Engineering and Manufacturing, 16*(5), 1035–1046.
2. Bikas, H., Stavropoulos, P., & Chryssolouris, G. (2016). Additive manufacturing methods and modelling approaches: A critical review. *The International Journal of Advanced Manufacturing Technology, 83*(1–4), 389–405.
3. Ngo, T. D., Kashani, A., Imbalzano, G., Nguyen, K. T., & Hui, D. (2018). Additive manufacturing (3D printing): A review of materials, methods, applications and challenges. *Composites Part B: Engineering, 143,* 172–196.
4. Singh, S., Ramakrishna, S., & Singh, R. (2017). Material issues in additive manufacturing: A review. *Journal of Manufacturing Processes, 25,* 185–200.
5. Rezaie, H. R., Esnaashary, M., Abolfazl Aref arjmand, & Öchsner, A. (2018). *A review of biomaterials and their applications in drug delivery.* Springer.
6. Khaing, M. W., Fuh, J. Y. H., & Lu, L. (2001). Direct metal laser sintering for rapid tooling: Processing and characterisation of EOS parts. *Journal of Materials Processing Technology, 113*(1–3), 269–272.
7. Alhnan, M. A., Okwuosa, T. C., Sadia, M., Wan, K. W., Ahmed, W., & Arafat, B. (2016). Emergence of 3D printed dosage forms: Opportunities and challenges. *Pharmaceutical Research, 33*(8), 1817–1832.
8. Rodriguez-Salvador, M., & Ruiz-Cantu, L. (2018). Revealing emerging science and technology research for dentistry applications of 3D bioprinting. *International Journal of Bioprint, 5*(1), 170.
9. Bhargav, A., Sanjairaj, V., Rosa, V., Feng, L. W., & Fuh YH, J. (2018). Applications of additive manufacturing in dentistry: A review. *Journal of Biomedical Materials Research Part B: Applied Biomaterials, 106*(5), 2058–2064.
10. Ligon, S. C., Liska, R., Stampfl, J., Gurr, M., & Mülhaupt, R. (2017). Polymers for 3D printing and customized additive manufacturing. *Chemical Reviews, 117*(15), 10212–10290.
11. Kruth, J. P., Vandenbroucke, B., Van Vaerenbergh, J., & Naert, I. (2005). Digital manufacturing of biocompatible metal frameworks for complex dental prostheses by means of SLS/SLM (pp. 139–146). Proceedings of VRAP, Leiria.

Chapter 7
Application of the Finite Element Method in Dentistry

7.1 Introduction

The finite element method (FEM) is a numerical method in which stress and strain distributions, resulting from the application of external force, pressure, and thermal variation, are calculated within several finite elements of a larger system with a complex geometry and heterogeneous properties [1]. Finite element method is, besides the strain gauge technique and the photoelastic test, the method frequently used in medical and dental applications to analyze stress and strain distributions within structures with complex shapes. The strain gauge method has limitations in the analysis of internal stresses; however, the photoelastic method is able to analyze the internal stress but it is difficult to build a model with satisfied accuracy in this method [2]. FEM permits to analyze and predict mechanical conditions or stress and strain distributions of biomaterials such as in teeth, dentures, implants, restoration materials, and neighboring bone, which are not available and accessible due to the complex biomechanical system of the oral cavity [3].

The FEM was first theoretically introduced by Turner in 1956. From then, this method was widely used in analyzing mechanical and structural problems [2]. Courant was the first, who utilized this method in biomechanical systems for analyzing biomechanical problems using numerical approximations of a discretized model and differential equations. Later in 1976, Weinstein used the FEM to analyze occlusion loading on implants and bones [1]. Ledley and Huang in 1968 were the first researchers who used the FEM in dentistry by using experimental data. Their work had an important rough simplification, which was conducting their model as a homogeneous structure. In fact, natural teeth are largely inhomogeneous with very different elastic moduli of the enamel and the dentin. Later, Farah, Thresher, and Yettram used the FEM for modeling the stress distribution within a tooth structure. Afterward, several researchers have been studied the stress distribution within teeth structures, restorative materials, implants, etc. Nowadays, the FEM is among

H. Reza Rezaie et al., *A Review on Dental Materials*, Advanced Structured Materials 123, https://doi.org/10.1007/978-3-030-48931-1_7

Table 7.1 Different advantages and disadvantages of FEM [1]

Advantages of FEM	Disadvantages of FEM
• Non-invasive • Reproducible • Low cost • Less time-consuming than other analyzing techniques • Able to simulate all mechanical and biological states • Useful for linear, nonlinear, static, and dynamic problems	• Sensitive to the accuracy of data input • Complexity of modeling human organs for each case • Requiring accurate values of the physical properties of the analyzed structure • Development of errors due to the various applied assumptions

the successful numerical methods used in medical and dental research fields [2]. Recently, the FEM is used to better simulate the conditions and properties of the oral environment by utilizing nonlinear models [4]. Different researchers have used the FEM in orthodontics, dentofacial orthopedics, operative dentistry, endodontics, prosthodontia, implantology, and for discovering the best geometry of fillings and restorative materials [1].

In this method, an approximation of the solution is obtained by the construction of a computer model of the structure. The model is physically discretized into several smaller and simpler parts (the so-called finite elements) with a finite number of degrees of freedom (DOF), defined dimensions, and defined physical properties. These elements are connected to each other via a finite number of points called nodes. The behavior of each element of the discretized system is calculated and together these elements can predict and simulate the behavior of the entire initial system. Shape and number of these finite elements are important factors influencing the achieved accuracy of the final state of the system. The power of the finite element analysis (FEA) is its adaptability, which can provide useful quantitative and visionary data at any location of the constructed model, which is not possible to be measured using conventional experimental methods [3]. Table 7.1 presents the various advantages and disadvantages of the FEM.

In this chapter, our main objective is to introduce the FEM, its advantages and disadvantages as well as to review the various applications of FEM in dentistry.

7.2 Finite Element Analysis Procedure

The finite element analysis (FEA) procedure consists of three-phase steps, which are the (I) pre-processing phase, (II) solution (processing) phase, and (III) post-processing phase. Several steps of each phase of FEA are presented in Table 7.2 [1].

Table 7.2 Different phases of finite element analysis besides their related objectives and steps [1, 5]

Phase of FEA	Phase objective	Steps
Pre-processing phase	Construction of the model, which is consisted of the geometrical representation, material properties, and boundary conditions	• Defining element type regarding the desired accuracy • Entering data such as material property (Young's modulus, poison's ratio, etc.) • Creating model using 3-D CT scan and 3-D laser scan • Mesh defining according to the complexity of model, time limitations, and desired accuracy
Solution (processing) phase	Calculating the desired result by the computer software	• Defining boundary conditions by restricting some degrees of freedom
Post-processing phase	Demonstrating the results, verifications, and conclusions	• Obtaining graphical, numerical, and animated result output

Frequently used commercial FEA packages in the field of dentistry are ABAQUS, ANSYS, and HYPER MESH [1].

The accuracy of the result output is dependent on the accuracy of the defined geometry, material properties, loads, and boundary conditions. The geometry of the model should represent the geometry of the actual system as exactly as possible. Two-dimensional finite element models save more time and costs than three-dimensional models due to their simple construction steps; however, they cannot accurately present the complex geometries or material properties of the model. On the other hand, three-dimensional models are more applicable than two-dimensional models for simulating complex geometries [3]. Three-dimensional models are capable of better visualizing the internal areas. These models require mesh refinement and are time-consuming [5]. Dry skull measurement and three-dimensional coordinate measuring are the conventional methods used for determining the geometrical shape of the dental materials [2]. To obtain and draw the most accurate model in the FEA software, digital imaging techniques such as computed tomography (CT) and magnetic resonance imaging (MRI) have been developed [3].

CT not only provides the morphology of bone and dental materials, but also determines their density in a living organ [2]. Computer-aided design (CAD) and Bio-CAD software are the techniques that can be used to construct 2D and 3D models of complex biological structures [5]. Nowadays, the detailed realistic models of complex geometrical shapes and layers of teeth are obtained from modern scanning procedures such as micro-XMT and micro-CT [6].

Defining suitable physical and mechanical properties to the various parts of the model is one of the challenging issues affecting the reliability and accuracy of the resulting output. Materials can be modeled as isotropic, anisotropic, orthotropic, hyperelastic, elastic, viscoelastic, plastic, etc. Dentistry materials are commonly modeled as homogenous and linearly isotropic materials due to the simplicity of model and related calculations [3]. In different studies, researchers have employed different properties to define the properties of dental structures. In this manner, some have allocated isotropic properties to enamel, which does not accurately simulate its properties. On the other hand, the dentin structure presents acceptable response while considered as isotropic structures. Most of the researches have been carried out using linear static models; however, these models are not completely valid. Actually, most of the real materials, contacts, and geometries present nonlinear characteristics. Nonlinear analysis provides more accurate results of the natural structures than linear analysis. In the field of dentistry, nonlinear finite element analysis has been used in various studies such as the simulation of periodontal ligament, tooth-to-tooth contact, and materials possessing plastic and viscoelastic behavior [5]. As bone, soft tissues and teeth are non-homogenous and anisotropic materials, the simplification in FEM may alter the results from their actual state; however, by taking caution and knowing the benefits and limitations of FEM, this method can bring about great advantages in the analysis of the function and longevity of dental materials within the oral environment [3].

Occlusal loads, which act on dental materials, are comprised of both axial and horizontal forces generating compressive and tensile stresses, respectively. Contrary to the tensile force, existence of compressive force is essential. The magnitudes of these forces are directly related to the region of mouth, bone shape, shape and diameters of the restoration, age, sex, muscle size, and so many other factors. Hence, the values are not defined exactly. Masticatory forces acting on dental materials and teeth are dynamic loads; however, to facilitate the calculating step of simulations, applied loads are assumed to be static. Boundary conditions, the last factor of the accuracy of results output are applied on the model to physically constraint it on relevant nodes. Physical constraints including displacements and supports ensure an equilibrium solution of the finite elements and must be applied on nodes to prevent displacement or rotation in the specific directions [3].

7.3 Application of the FEM in Dentistry

Table 7.3 summarizes various studies and their results regarding the application of FEA in the field of dentistry.

Table 7.3 Various applications of the FEM in the field of dentistry [1–3, 5–10]

Dentistry specialty	Researchers	Model	Object	Results
Orthodontics & dentofacial orthopedics	Liang et al. (2009)	3D finite element models of maxilla and maxillary incisors	Evaluation of the torque control of maxillary incisors in labial and lingual orthodontic technique	Stress & strain distribution on periodontal ligament and adjacent accompanied with movements of incisors were studied. Results demonstrated similar forces in bodily movement in labia technique and uncontrolled tipping in lingual technique
	Field et al. (2009)	Two 3D finite element models of single mandibular canine and mandibular incisor, canine and premolars	Evaluation of the stress distribution in single- and multi-tooth system	The crest of alveolar ridge showed an increased distortion strain and the tooth apex presented tensile and compressive stresses causing root resorption. Multi-tooth system showed higher levels of stress compared to single tooth system
	Magne et al. (2002)	2D finite element models of maxillary and mandibular molar teeth during the working, nonworking, and vertical closure of the jaw using nonlinear contact analysis	Analysis of the contact and stress distribution of posterior tooth surfaces	During nonworking motion, high-stress concentrations were generated in the central groove of the maxillary tooth. During working closure, stress concentrations were formed on the lingual surface of the mandibular tooth

(continued)

Table 7.3 (continued)

Dentistry specialty	Researchers	Model	Object	Results
	Rudolph et al. (2001)	3D finite element model of a maxillary central incisor, periodontal ligament, and alveolar bone	Evaluation of the stress distribution at root apex during orthodontic tooth movements	During intrusive, extrusive, and rotational movements, greater stresses were developed at apex. Tipping force caused generation of stress at the alveolar crest. During bodily movement, stress developed within the periodontal ligament
	Dejak et al. (2003)	2D finite element model the mandibular molar teeth with PDL	Analysis of stress distribution in molar teeth during clenching and mastication	Food pieces with low rigidity provide great values of stress in the cervical region of the lingual alveolar bone of the mandibular molar tooth, whereas nonhomogeneous food pieces containing fragments of bone create high stress in the enamel surface
	Hasegawa et al. (2010)	3D finite element model of the lower first premolar (with the three layers of enamel, dentin, and pulp) and the mandible (with the two layers of cortical and cancellous bones) using CT images	Analyzing the stresses on the teeth and supporting jawbone structure during occlusion and evaluate the possibility and reliability of mechanical simulation	The possibility of performing a finite element model with high accuracy using X-ray CT images was shown in this study. Stress was concentrated in the buccal cervical enamel and adjacent bone, which can cause wedge-shaped defects that are believed to be induced due to the repeated stress of occlusal forces

(continued)

Table 7.3 (continued)

Dentistry specialty	Researchers	Model	Object	Results
Restorative dentistry and Endodontics	Ichim et al. (2007)	3D finite element model of cervical glass ionomer restoration	Assessing the co-relation of cavity shape, depth and occlusal forces with durability of GIC restoration	Depth and shape of the GIC restoration did not present any important effect on the durability of restoration. Readjustment of the inter-occlusal contacts can provide higher retention of restoration
	Asmussen et al. (2008)	3D finite element models of class I and class II restorations	Assessing the influence of modulus of elasticity of resin composite on stresses generated by occlusal loading to reduce marginal failure	Restorations possessing high modulus of elasticity present lower risk of marginal deterioration
	Yamamoto et al. (2007)		Analysis of indirect restorations prepared in cavity bases	Materials possessing high modulus of elasticity are more suitable cavity bases for posterior restorations
	Subramaniam et al. (2007)	3D finite element model of rotary nickel–titanium instruments	Evaluating the bending & torsional stress distributions in Protaper & Profile nickel–titanium rotary instruments	Applying equal loads to Protaper and Profile models provide uniform stress distribution within the Protaper and less elasticity in comparison with the Profile model
	Hong et al. (2003)	3D finite element model of the root canal wall	Developing the best method of condensation in root canal to avoid its vertical fracture	Vertical condensation method brings about high stress on root canal walls compared to lateral condensation

(continued)

Table 7.3 (continued)

Dentistry specialty	Researchers	Model	Object	Results
	Belli et al. (2005) and Campos et al. (2011)		Evaluation of the stress distribution with the use of ceramics and polymers in indirect restorations	Ceramic indirect restorations present more favorable characteristics than polymers due to the less stress transfer through the tooth structure
	Ausiello et al. [6]	3D finite element model of indirect composite-restored molar tooth with class II MOD cavity using a CAD–FEM procedure, fatigue mechanic laws, and linear analysis	Evaluation of the fatigue lifetimes before crack failure of a restored tooth	Highest static stress was located around the fillets of the class II MOD cavity. These locations demonstrated the shortest fatigue-life
	Ausiello et al. [8]	One 3D model of a sound lower molar and three 3D class II MOD cavity models with 95° cavity-margin-angle shapes were modeled with different material combinations using micro-CT scanner system and linear analysis of isotropic materials	Evaluation of the influence of resin based and lithium disilicate material combinations on the stress and strain distributions in adhesive class II mesio-occlusal-distal (MOD) restorations during mastication and their ability to relieve stresses	Bulk filling resin composite with 1% polymerization shrinkage caused marginal and internal stress in large Class II MOD adhesive restorations. Machinable resin block composite and lithium disilicate adhesive inlay restorations performed favorably under loading and shrinkage. Both inlay systems absorbed stress comparably to a sound tooth. Shrinkage stress peaks were mainly located marginally along the enamel-restoration interface at occlusal and mesiodistal sites

(continued)

Table 7.3 (continued)

Dentistry specialty	Researchers	Model	Object	Results
	Suzuki et al. (2008) and Soares et al. (2010)		Investigation of stress distribution in roots restored with different luting agents	Resin cement show higher fracture toughness compared to other cements. Besides this, zinc–phosphate cement and conventional glass ionomer cement cause greater levels of stress concentration at the interface of cement and dentin
Prosthodontic & Implantology	Bramanti et al. (2017)	3D finite element models of single crown prosthodontics of chrome cobalt alloy, golden alloy, dental resin, and zirconia using linear static analysis	Evaluation of the stress distribution while applying different single crown prosthodontic materials over masticatory cycle	Dental resin provided low value of fracture whereas metal alloy and zirconia have presented high values of fracture. Zirconia is the best substructure for anterior free contact teeth. On the other hand, chrome cobalt alloy acts better as a posterior restorative material
	Boschain et al. (2006)	3D finite element model of a post-restored tooth	Evaluation of the stress distribution in a post-restored tooth	Post materials possessing an elastic modulus higher than dentin can cause failure and non-homogenous stresses in root dentin. The best performance is achieved by using posts, which have similar stiffness to dentin

(continued)

Table 7.3 (continued)

Dentistry specialty	Researchers	Model	Object	Results
	Eraslan et al. (2009)	3D finite element model of cervical ferule and post-and-core-restored all-ceramic anterior crowns assumed as homogenous and isotropic materials	Evaluation of the effect of cervical ferrule height on stress distribution at post-and-core-restored all-ceramic anterior crowns	Utilization of a cervical ferrule reduces stress concentration in endodontically treated teeth, which are restored with an all-ceramic post-and-core. The stress concentration within the zirconium oxide post was higher than fiber posts
	Siegele et al. (1989)	3D finite element model of implant with different shapes inserted in the jawbone	Evaluating the stress distribution through the jawbone after insertion of implants with different shapes	Implant shape is directly associated with the stress distribution in the jawbone. Conical implant, for example, generates higher stresses than cylindrical and screw-shaped implants
	Himmlova et al. (2004)	3D finite element model of implants with various lengths and diameters	Evaluation of the stress distribution at the implant-bone interface	Neck of implant presents maximum stress concentration. The stress decreases with increase in diameter of implant (up to 4.5 mm) and increase in length of implant. Diameter of implant presents much more influence in stress generation than its length

(continued)

Table 7.3 (continued)

Dentistry specialty	Researchers	Model	Object	Results
	Dos Santos et al. (2011)	3D finite element model of peri-implant bone with relined dentures and different heights of healing caps	Evaluation of the stress distribution of submerged & non-submerged implants	Non-submerged implants present increased stress concentration than submerged implants and soft liner materials are better choices to be used than the not relined denture bases. The height of the healing caps seems is an important factor in stress formation in the peri-implant bone through the healing period
	Erić et al. (2016)	3D finite element models of four-unit tooth–tooth, implant–implant and tooth–implant supported dentures and the supporting tissues using ATOS scanner and CATIA	Evaluation of stress distributions in supporting tissues in four-unit tooth–tooth, implant–implant and tooth–implant supported dentures	Applying occlusal forces generates maximum stress concentration at the neck of implant and tooth in all models, whereas applying axial and oblique loading on the pontic generates the maximum stress concentrations at the implant neck for tooth–implant supported fixed partial denture. Implant–implant supported fixed partial denture show better results compared to tooth–implant fixture

(continued)

Table 7.3 (continued)

Dentistry specialty	Researchers	Model	Object	Results
	Glišić et al. (2016)	Two 3D finite element models of two types of implants mounted in the second molar position associated with tooth contours, periodontal ligament, mucous membranes, cortical bones and spongiosa, abutment and suprastructure. All materials except periodontal ligament were assumed homogenous, linear elastic, and isotropic	Analysis of stress distribution and deformation in tooth–implant supported fixed partial dentures and surrounding tissue with tooth–implant connection for two types of implants (resilient Titan Shock Absorber (TSA) abutment and conventional non-resilient abutment)	Resilient TSA abutment presented lower stress values compared to non-resilient abutment
	Grbović et al. (2009)	3D finite element model of full denture with two mini dental implants	Optimization of the number of mini dental implants (MDI) to retain complete denture while alveolar ridge is poorly developed	Models of the denture supported by only alveolar ridge presented great non-stability whereas denture supported by MDIs were stable and MDIs present substantial support for lower complete denture

(continued)

Table 7.3 (continued)

Dentistry specialty	Researchers	Model	Object	Results
	El-Anwar and El-Zawahry [10]	3D finite element models of 25 implants with gradual increase in diameter and length using linear static analysis	Evaluation of the stress distribution in spongy and cortical bones effect while using different geometric configurations of implants	Increasing implant diameter and length provides better stress distribution on spongy and cortical bones. The behavior of wide implants (5 and 6 mm diameter) was not greatly affected when its length was increased. On the other hand, the behavior of small diameter implants was gradually improved by increasing implant length. Consequently, implant length is an important parameter in case of using implants with small diameters

References

1. Mohammed, S. D., & Desai, H. (2014). Basic concepts of finite element analysis and its applications in dentistry, an overview. *Oral Hyg Health, 2*(156), 2332-0702.
2. Hasegawa, A., Shinya, A., Nakasone, Y., Lassila, L. V., Vallittu, P. K., & Shinya, A. (2010). Development of 3D CADFEM analysis system for natural teeth and jaw bone constructed from X-ray CT images. Interna
3. Grbović, A., & Mihajlović, D. (2017). Practical aspects of finite element method applications in dentistry. *Balkan Journal of Dental Medicine, 21*(2), 69–77.
4. Shetty, P., Hegde, A., & Rai, K. (2010). Finite element method—an effective research tool for dentistry. *Journal of Clinical Pediatric Dentistry, 34*(3), 281–285.
5. Piccioni, M. A. R., Campos, E. A., Saad, J. R. C., de Andrade, M. F., Galvão, M. R., & Rached, A. A. (2013). *Application of the finite element method in Dentistry*. RSBO Revista Sul-Brasileira de Odontologia
6. Ausiello, P., Franciosa, P., Martorelli, M., & Watts, D. C. (2011). Numerical fatigue 3D-FE modeling of indirect composite-restored posterior teeth. *Dental Materials, 27*(5), 423–430.
7. Wakabayashi, N., Ona, M., Suzuki, T., & Igarashi, Y. (2008). Nonlinear finite element analyses advances and challenges in dental applications. *Journal of Dentistry, 36*(7), 463–471.
8. Ausiello, P., Ciaramella, S., Fabianelli, A., Gloria, A., Martorelli, M., Lanzotti, A., & Watts, D. C. (2017). Mechanical behavior of bulk direct composite versus block composite and lithium disilicate indi
9. Bramanti, E., Cervino, G., Lauritano, F., Fiorillo, L., D'Amico, C., Sambataro, S., & Cicciù, M. (2017). FEM and von mises analysis on prosthetic crowns structural elements Evaluation of different appli
10. El-Anwar, M. I., & El-Zawahry, M. M. (2011). A three dimensional finite element study on dental implant design. *Journal of Genetic Engineering and Biotechnology, 9*(1), 77–82.

The manufacturer's authorised representative in the EU is Springer Nature Customer Service Centre GmbH, Europaplatz 3, 69115 Heidelberg, Germany. If you have any concerns regarding our products, please contact ProductSafety@springernature.com

Printed and bound by CPI Group (UK) Ltd, Croydon, CR0 4YY
30/09/2026
02225559-0008